Greener Catalysis for Environmental Applications

Greener Catalysis for Environmental Applications

Editor

Stanisław Wacławek

MDPI • Basel • Beijing • Wuhan • Barcelona • Belgrade • Manchester • Tokyo • Cluj • Tianjin

Editor
Stanisław Wacławek
Institute for Nanomaterials,
Advanced Technologies and
Innovation
Technical University of Liberec
Liberec
Czech Republic

Editorial Office
MDPI
St. Alban-Anlage 66
4052 Basel, Switzerland

This is a reprint of articles from the Special Issue published online in the open access journal *Catalysts* (ISSN 2073-4344) (available at: www.mdpi.com/journal/catalysts/special_issues/green_env).

For citation purposes, cite each article independently as indicated on the article page online and as indicated below:

LastName, A.A.; LastName, B.B.; LastName, C.C. Article Title. *Journal Name* **Year**, *Volume Number*, Page Range.

ISBN 978-3-0365-3978-2 (Hbk)
ISBN 978-3-0365-3977-5 (PDF)

© 2022 by the authors. Articles in this book are Open Access and distributed under the Creative Commons Attribution (CC BY) license, which allows users to download, copy and build upon published articles, as long as the author and publisher are properly credited, which ensures maximum dissemination and a wider impact of our publications.

The book as a whole is distributed by MDPI under the terms and conditions of the Creative Commons license CC BY-NC-ND.

Contents

About the Editor vii

Preface to "Greener Catalysis for Environmental Applications" ix

Stanisław Wacławek
Greener Catalysis for Environmental Applications
Reprinted from: *Catalysts* 2021, *11*, 585, doi:10.3390/catal11050585 1

Kamil Krawczyk, Stanisław Wacławek, Edyta Kudlek, Daniele Silvestri, Tomasz Kukulski and Klaudiusz Grübel et al.
UV-Catalyzed Persulfate Oxidation of an Anthraquinone Based Dye
Reprinted from: *Catalysts* 2020, *10*, 456, doi:10.3390/catal10040456 5

Edyta Kudlek
Transformation of Contaminants of Emerging Concern (CECs) during UV-Catalyzed Processes Assisted by Chlorine
Reprinted from: *Catalysts* 2020, *10*, 1432, doi:10.3390/catal10121432 21

Maria Kurańska and Magdalena Niemiec
Cleaner Production of Epoxidized Cooking Oil Using A Heterogeneous Catalyst
Reprinted from: *Catalysts* 2020, *10*, 1261, doi:10.3390/catal10111261 39

Alamri Rahmah Dhahawi Ahmad, Saifullahi Shehu Imam, Wen Da Oh and Rohana Adnan
Fe_3O_4-Zeolite Hybrid Material as Hetero-Fenton Catalyst for Enhanced Degradation of Aqueous Ofloxacin Solution
Reprinted from: *Catalysts* 2020, *10*, 1241, doi:10.3390/catal10111241 53

Hsu-Hui Cheng, Shiao-Shing Chen, Hui-Ming Liu, Liang-Wei Jang and Shu-Yuan Chang
Glycine–Nitrate Combustion Synthesis of Cu-Based Nanoparticles for NP9EO Degradation Applications
Reprinted from: *Catalysts* 2020, *10*, 1061, doi:10.3390/catal10091061 73

Michael Meistelman, Dan Meyerstein, Amos Bardea, Ariela Burg, Dror Shamir and Yael Albo
Reductive Dechlorination of Chloroacetamides with $NaBH_4$ Catalyzed by Zero Valent Iron, ZVI, Nanoparticles in ORMOSIL Matrices Prepared via the Sol-Gel Route
Reprinted from: *Catalysts* 2020, *10*, 986, doi:10.3390/catal10090986 87

Esneyder Puello-Polo, Yina Pájaro and Edgar Márquez
Effect of the Gallium and Vanadium on the Dibenzothiophene Hydrodesulfurization and Naphthalene Hydrogenation Activities Using Sulfided NiMo-V_2O_5/Al_2O_3-Ga_2O_3
Reprinted from: *Catalysts* 2020, *10*, 894, doi:10.3390/catal10080894 105

Sonia Bonacci, Giuseppe Iriti, Stefano Mancuso, Paolo Novelli, Rosina Paonessa and Sofia Tallarico et al.
Montmorillonite K10: An Efficient Organo-Heterogeneous Catalyst for Synthesis of Benzimidazole Derivatives
Reprinted from: *Catalysts* 2020, *10*, 845, doi:10.3390/catal10080845 125

Zahra Gholami, Zdeněk Tišler, Romana Velvarská and Jaroslav Kocík
CoMn Catalysts Derived from Hydrotalcite-Like Precursors for Direct Conversion of Syngas to Fuel Range Hydrocarbons
Reprinted from: *Catalysts* 2020, *10*, 813, doi:10.3390/catal10080813 137

Constantine Tsounis, Yuan Wang, Hamidreza Arandiyan, Roong Jien Wong, Cui Ying Toe and Rose Amal et al.
Tuning the Selectivity of LaNiO$_3$ Perovskites for CO$_2$ Hydrogenation through Potassium Substitution
Reprinted from: *Catalysts* **2020**, *10*, 409, doi:10.3390/catal10040409 . . . . **155**

About the Editor

Stanisław Wacławek

Stan, during his scientific career, visited, among others, the University of Guelph (Canada) and the University of Cincinnati (USA), where he conducted research on sulfate radical-based processes. He is in 2% of the most-cited scientists in the world, according to Scopus and a Publons 2019 awardee: the top 1% of reviewers in Environment and Cross-Field. He serves on the editorial board of, e.g., {Chemical Engineering Journal} (IF: 13.273) and {Catalysts} (IF: 4.146). In 2022 he starts the Associate Professor position at the Technical University of Liberec.

Preface to "Greener Catalysis for Environmental Applications"

This reprint is focused on the greener catalytic reactions for various environmental applications. I want to thank all the authors involved in this work. Their publications are of high scientific value and are highly appreciated.

I would also like to dedicate this book to four wonderful people in my life: my Mother - Maria, Wife - Aneta, Daughter - Misia, and Son - Maksymilian. I would like to thank my mother and wife for their amazing inspiration, help and patience.

Stanisław Wacławek
Editor

Editorial
Greener Catalysis for Environmental Applications

Stanisław Wacławek

Institute for Nanomaterials, Advanced Technologies and Innovation, Technical University of Liberec, Studentská 1402/2, 461 17 Liberec, Czech Republic; stanislaw.waclawek@tul.cz; Tel.: +420-485-353-006

Catalytic reactions account for approximately 85% of all chemical reactions, and they are particularly significant in environmental science. Anastas and Warner introduced their 12 postulates of green chemistry, the ninth of which is catalysis, more than 20 years ago. The catalysts can be further made and used in such a way that the environmental benefits could be even more.

This Special Issue is devoted to "greener catalysis for environmental applications", and primarily covers the catalytic synthesis of value-added chemicals, as well as the catalytic removal of pollutants.

One of the examples of catalytic removal of contaminants in water by the activation of an oxidant was presented by Krawczyk et al. [1]. The mechanism of decolorization of Acid Blue 129 was explained experimentally and theoretically, and the toxicity decline (evaluated using *Daphnia magna* and *Lemna minor*) of the solution after the oxidation was observed. However, this is not always the case with advanced oxidation processes, as noted by Kudlek [2]. The catalytically activated oxidants, and the radicals created in that process, can generate disinfection byproducts, which are often substantially more toxic compared to the original substrate (e.g., case of ibuprofen in the article of Kudlek [2]). Nonetheless, after a sufficient treatment period, UV-catalyzed processes may decrease the toxic nature of post-processed water solutions. The biodegradation of some compounds, such as nonylphenol ethoxylates (NPEOs), can also lead to toxic intermediates, which further require more invasive treatment. The heterogeneous catalytic activation of hydrogen peroxide was found to be a very efficient system for the decomposition of NPEOs, with an average of nine ethylene oxide units (NP9EO) [3]. A nanocrystalline Cu-based heterogeneous catalyst, in a dose of 0.3 g/L and an H_2O_2 concentration of 0.05 mM, has resulted in NP9EO and total organic carbon (TOC) removal efficiency of 83.1% and 70.6%, respectively. Other catalysts for the heterogeneous activation of hydrogen peroxide are iron materials such as magnetite. Zeolite-supported magnetite was used to activate H_2O_2 for the first time for the removal of ofloxacin [4]. This system achieved 88% of ofloxacin degradation efficiency and 51% of TOC removal efficiency under optimized reaction conditions. Furthermore, after five runs, reusability tests showed only a slight decrease in the catalytic activity.

The catalyst's reusability was also presented in work focused on the reductive removal of pollutants [5]. Therein the authors have utilized elemental iron as a catalyst for the reduction of chloroacetamides. By varying the amount of catalyst or reducing agent before the reaction, it was possible to obtain conditions for the complete dechlorination of these pollutants to nontoxic substances. The reductive treatment of dibenzothiophene and naphthalene was presented by Puello-Polo et al. [6]. They have determined that the addition of gallium and vanadium as structural promoters in the NiMo/Al_2O_3 catalysts allows for the largest generation of sites for the hydrogenation and desulfurization of contaminants.

Greener removal of contaminants can be turned even more sustainable by pairing it with the simultaneous synthesis of value-added products. The products of complete and incomplete combustion of hydrocarbons, i.e., carbon dioxide (CO_2) and carbon monoxide (CO), are considered pollutants harmful to humankind and the environment. In such a sense, authors [7] have demonstrated the substitution of La with K cations in $LaNiO_3$

perovskite that exhibited a 100% selectivity towards the methanation of CO_2 at all temperatures investigated. On the other hand, reducing CO to value-added products such as gasoline and jet fuel range hydrocarbons by two different groups of CoMn catalysts derived from hydrotalcite-like precursors was reported by Gholami et al. [8]. The catalysts prepared using a KOH + K_2CO_3 mixture as a precipitant agent exhibited a high selectivity of 51–61% for gasoline (C_5–C_{10}) and 30–50% for jet fuel (C_8–C_{16}) range hydrocarbons compared with catalysts precipitated by KOH.

In a typical batch chemical process, solvents account for fifty to eighty percent of the mass, and they also drive its majority energy consumption. In this regard, for the greener synthesis of benzimidazole derivatives, the authors have reported solvent-free conditions and short synthesis time by a green montmorillonite K10 catalyzed method [9]. They have claimed that this method does not require the use of solvents, and can substantially reduce energy consumption in comparison to recently published procedures. Similarly, a greener solvent-free process of used cooking oil epoxidation has been developed [10]. An additional advantage of this method can be the catalyst's enhanced activity after reusing (for not more than four times).

In conclusion, this Special Issue gathered articles of substantial quality and broad scientific interest to the *Catalysts* research community on various ways for making catalytic processes greener. Catalysis sustainability improvement can be obtained, among other things, by careful toxicity assessment during the catalyst preparation and the use of solvent-free conditions, catalyst recyclability, and pairing the removal of contaminants with the synthesis of value-added products (Figure 1).

Figure 1. Scheme presenting the main topics that concern greener catalysis and were addressed in this Special Issue.

Funding: This research received no external funding.

Acknowledgments: The guest editor would like to express his gratitude to the authors who have contributed to this Special Issue.

Conflicts of Interest: The author declares no conflict of interest.

References

1. Krawczyk, K.; Wacławek, S.; Kudlek, E.; Silvestri, D.; Kukulski, T.; Grübel, K.; Padil, V.V.T.; Černík, M. UV-catalyzed persulfate oxidation of an anthraquinone based dye. *Catalysts* **2020**, *10*, 456. [CrossRef]
2. Kudlek, E. Transformation of contaminants of emerging concern (CECs) during UV-catalyzed processes assisted by chlorine. *Catalysts* **2020**, *10*, 1432. [CrossRef]

3. Cheng, H.H.; Chen, S.S.; Liu, H.M.; Jang, L.W.; Chang, S.Y. Glycine–nitrate combustion synthesis of Cu-based nanoparticles for NP9EO degradation applications. *Catalysts* **2020**, *10*, 1061. [CrossRef]
4. Ahmad, A.R.D.; Imam, S.S.; Oh, W.D.; Adnan, R. Fe_3O_4-zeolite hybrid material as hetero-fenton catalyst for enhanced degradation of aqueous ofloxacin solution. *Catalysts* **2020**, *10*, 1241. [CrossRef]
5. Meistelman, M.; Meyerstein, D.; Bardea, A.; Burg, A.; Shamir, D.; Albo, Y. Reductive dechlorination of chloroacetamides with $NaBH_4$ catalyzed by zero valent iron, ZVI, nanoparticles in ORMOSIL matrices prepared via the sol-gel route. *Catalysts* **2020**, *10*, 986. [CrossRef]
6. Puello-Polo, E.; Pájaro, Y.; Márquez, E. Effect of the gallium and vanadium on the dibenzothiophene hydrodesulfurization and naphthalene hydrogenation activities using sulfided nimo-V_2O_5/Al_2O_3-Ga_2O_3. *Catalysts* **2020**, *10*, 894. [CrossRef]
7. Tsounis, C.; Wang, Y.; Arandiyan, H.; Wong, R.J.; Toe, C.Y.; Amal, R.; Scott, J. Tuning the Selectivity of $LaNiO_3$ Perovskites for CO_2 hydrogenation through potassium substitution. *Catalysts* **2020**, *10*, 409. [CrossRef]
8. Gholami, Z.; Tišler, Z.; Velvarská, R.; Kocík, J. Comn catalysts derived from hydrotalcite-like precursors for direct conversion of syngas to fuel range hydrocarbons. *Catalysts* **2020**, *10*, 813. [CrossRef]
9. Bonacci, S.; Iriti, G.; Mancuso, S.; Novelli, P.; Paonessa, R.; Tallarico, S.; Nardi, M. Montmorillonite K10: An efficient organo-heterogeneous catalyst for synthesis of benzimidazole derivatives. *Catalysts* **2020**, *10*, 845. [CrossRef]
10. Kurańska, M.; Niemiec, M. Cleaner production of epoxidized cooking oil using a heterogeneous catalyst. *Catalysts* **2020**, *10*, 1261. [CrossRef]

Article

UV-Catalyzed Persulfate Oxidation of an Anthraquinone Based Dye

Kamil Krawczyk [1], Stanisław Wacławek [1,*], Edyta Kudlek [2], Daniele Silvestri [1,*], Tomasz Kukulski [3], Klaudiusz Grübel [4], Vinod V. T. Padil [1] and Miroslav Černík [1]

1. Institute for Nanomaterials, Advanced Technologies and Innovation, Technical University of Liberec, Studentská 1402/2, 46117 Liberec 1, Czech Republic; kamil.krawczyk@tul.cz (K.K.); vinod.padil@tul.cz (V.V.T.P.); miroslav.cernik@tul.cz (M.Č.)
2. Department of Water and Wastewater Engineering, Silesian University of Technology, Konarskiego 18, 44-100 Gliwice, Poland; edyta.kudlek@polsl.pl
3. Institute of Textile Engineering and Polymer Materials, University of Bielsko-Biala, Willowa 2, 43-309 Bielsko-Biala, Poland; tkukulski@ath.bielsko.pl
4. Institute of Environmental Protection and Engineering, University of Bielsko-Biala, Willowa 2, 43-309 Bielsko-Biala, Poland; kgrubel@ath.bielsko.pl
* Correspondence: stanislaw.waclawek@tul.cz (S.W.); daniele.silvestri@tul.cz (D.S.)

Received: 15 March 2020; Accepted: 21 April 2020; Published: 23 April 2020

Abstract: Wastewater from the textile industry has a substantial impact on water quality. Synthetic dyes used in the textile production process are often discharged into water bodies as residues. Highly colored wastewater causes various of problems for the aquatic environment such as: reducing light penetration, inhibiting photosynthesis and being toxic to certain organisms. Since most dyes are resistant to biodegradation and are not completely removed by conventional methods (adsorption, coagulation-flocculation, activated sludge, membrane filtration) they persist in the environment. Advanced oxidation processes (AOPs) based on hydrogen peroxide (H_2O_2) have been proven to decolorize only some of the dyes from wastewater by photocatalysis. In this article, we compared two very different photocatalytic systems (UV/peroxydisulfate and UV/H_2O_2). Photocatalyzed activation of peroxydisulfate (PDS) generated sulfate radicals ($SO_4^{\bullet-}$), which reacted with the selected anthraquinone dye of concern, Acid Blue 129 (AB129). Various conditions, such as pH and concentration of PDS were applied, in order to obtain an effective decolorization effect, which was significantly better than in the case of hydroxyl radicals. The kinetics of the reaction followed a pseudo-first order model. The main reaction pathway was also proposed based on quantum chemical analysis. Moreover, the toxicity of the solution after treatment was evaluated using *Daphnia magna* and *Lemna minor*, and was found to be significantly lower compared to the toxicity of the initial dye.

Keywords: photocatalysis; dye; UV; peroxydisulfate; advanced oxidation process

1. Introduction

A source of clean water is important for various industrial, social and economic development sectors; therefore, it has to be constantly monitored for impurities. Increased human activity has introduced a wide range of toxic chemicals including inorganic (e.g., chromium, mercury, lead) and organic (e.g., pesticides, surfactants, pharmaceuticals) pollutants into the aqueous environment [1,2]. A significant source of such polluting compounds is wastewater from the textile industry, which is classified as the most polluting of all industrial sectors in terms of effluent volume and its chemical content [3]. The chemical loads of textile effluents originate from the residues of textile production processes, such as printing, scouring, bleaching and dyeing [4]. During the batch dyeing process, which is a common method for dying textiles, approximately 10% to 15% of the synthetic dyes used are lost, due to the inefficiency of the operation [5]. The residues are discharged into the effluent, from which they cannot be effectively removed by conventional wastewater treatment processes [6].

Dyes may be classified by their application or chemical structure into direct dyes (polyazo compounds, stilbenes, oxazines), basic dyes (diazahemicyanine, hemicyanine, cyanine, thiazine, acridine) or solvent dyes (azo, anthraquinone) [7]. Polyazo dyes have three or more N=N bonds in the molecule, and the number of azo groups attached to its center determines the color index of the dye (CI, systematic classification of colors by their saturation, brightness and hue) [8]. Designed to be highly stable towards light, temperature, water and detergents, dyes persist in the environment [9]. The presence of one or more benzene rings in their structure makes them more recalcitrant to biodegradation [10]. Moreover, dyes discharged into water even at a low concentration (even below 1 mg/L) are not only highly visible, which affects the aesthetic quality and transparency of water bodies (lakes, rivers) [11], but they also disturb the aquatic life by reducing light penetration and inhibiting photosynthesis, which causes oxygen deficiency [12]. Azo and anthraquinone dyes represent around 90% of all organic colorants [13]. They pose a threat to aquatic organisms (bacteria, algae, fish) by being toxic (lethal effect, genotoxic, mutagenic, carcinogenic) [14,15]. In particular, Acid Blue 129 (AB129), which is an acidic dye with an anthraquinone structure, being extensively used in the dyeing of silk, wool, cotton, paper, leather and nylon [16], was found to be associated with an ecotoxic hazard and danger of bioaccumulation [17]. The properties and structure of Acid blue 129 (AB129) are described in Table 1.

Table 1. Acid blue 129 (AB129) properties and structure

Chemical Formula: $C_{23}H_{19}N_2NaO_5S$	
Name: Sodium-1-amino-4-(2, 4, 6-trimethylanilino) anthraquinone-2-sulfonate	
Molecular weight: 458.46	
Melting point: >300 °C	Acid blue 129 structure

The conventional treatments of textile effluents involve, among others things: adsorption, coagulation-flocculation, membrane filtration and activated sludge [18,19]. However, these methods are not completely efficient and have several shortcomings. The adsorption process usually involves the use of activated carbon, which is expensive and incurs additional to regeneration and disposal costs [20]. Several dyes can inhibit bacteria development in activated sludge or cause membrane fouling using the filtration method [21,22]. Coagulation–flocculation is a pH-dependent process, which generates an extensive amount of concentrated sludge and is not suitable for all dyes [23].

Considering the obstacles in conventional textile wastewater treatment, alternative methods were developed. One of which is the advanced oxidation process (AOP), which utilizes highly reactive oxidizing intermediates like hydroxyl radicals (•OH) [24]. These radicals are often catalytically generated from hydrogen peroxide (H_2O_2) or ozone. For example, the Fenton reaction uses iron as a catalyst for producing •OH. AOPs can also utilize ozone (O_3) to produce •OH, which is used for decolorization of the azo dyes C.I. Reactive black 5 [25]. Ultraviolet (UV) radiation can catalyze the generation of •OH by photolysis of H_2O_2. This process was reported to be effective in the degradation of some dyes [26,27]. UV is also extensively used in combination with O_3 [28]. In one study, besides acting as a catalyst, UV radiation also contributed to the enhanced removal of total organic carbon (TOC) and chemical oxygen demand (COD), in a decolorization experiment of Reactive Blue 19 [29]. While the oxidation-reduction potential (ORP) of •OH/H_2O is 2.8 V (at an acidic pH) and •OH/OH⁻ is 1.89 V (at an alkaline pH) [30], the use of H_2O_2 to generate radicals is not cost-effective. For example, Argun and Karatas [31] reported that 4 g/L of H_2O_2 and 0.2 g/L of iron salt were used to decolorize 200 mg/L of synthetic dye. Similarly, Meric et al. [32] used 0.4 g/L of H_2O_2 and 0.1 g/L of iron salt to decolorize 100 mg/L of dye. Therefore, high consumption of H_2O_2 provides an economical challenge and increases the need to find cheaper and more effective substitutes, e.g., permanganate, ozone, persulfate anions or sulfate radicals [33]. Sulfate radicals ($SO_4^{•-}$) and •OH- based oxidation processes have comparable reaction rates for the removal of some pharmaceuticals [34], but sulfate radicals usually have a longer half-life (30–40 μs) than •OH (10^{-3} μs) [35,36]. Both radicals differ also in their reaction behavior, whereby $SO_4^{•-}$ favors electron transfer and •OH reacts more by addition and H-abstraction [37,38]. This is reflected in the different types of dyes treated. For example, Tang and Huren [39] reported that •OH is ineffective for the oxidation of anthraquinone dyes, while degradation by $SO_4^{•-}$ is effective [40]. $SO_4^{•-}$ may be generated by catalytic activation of peroxydisulfate (PDS) by: heat, UV radiation, transition metals, electrolysis, transition metals or radiolysis [41,42]. While PDS in the form of sodium persulfate is cheaper (0.74 USD/kg) [43] and safer to handle than liquid H_2O_2 (1.5 USD/kg), it is more expensive if calculated per mol (0.18 USD/mol PDS, 0.05 USD/mol H_2O_2), and hence the amount of radicals generated. Despite the many advantages of persulfate treatment, its disadvantages also have to be taken into consideration, such as post-treatment toxicity. Post-contamination with sulfate salts may be thought a small problem in comparison to the toxic by-products formed in a $SO_4^{•-}$ system, including transformation products of target contaminants (e.g., polynitrophenol compounds formed from nitrophenols [44]) and the by-products generated from effluent organic matter. $SO_4^{•-}$ is known to be more prone to form such post-contamination; therefore, toxicity studies after persulfate treatment are recommended [41].

In this study, photocatalyzed decolorization experiments of anthraquinone dye AB129 were conducted under various conditions. The work was performed to evaluate the role of sulfate and hydroxyl radicals in the dye oxidation. Pseudo-first order rate kinetics were also evaluated, and a simple pathway was proposed. Finally, the post-treatment toxicity of by-products was measured. To the best of our knowledge, this is the first time that the UV application of PDS has been used for the catalyzed oxidation of anthraquinone dye. Table 2 shows the various methods used to degrade anthraquinone dyes.

Table 2. Methods used to degrade anthraquinone dyes.

Method Used	Dye	Decolorization [%]/Time [h]	Reference
AOP (wet air, wet peroxide, Fenton, photocatalytic,)	Reactive Blue 4	100%, 100%, 99%/1 and 100%/0.75	[45]
AOP (TiO_2 and ZnO nanoparticles + photodegradation)	Reactive Blue 19	>95%/0.5	[46]
AOP (ozonation)	Reactive Blue 19	~100%/0.3	[47]
AOP (TiO_2 + photodegradation)	Reactive Blue 19	~75%/3	[48]
AOP (Fenton reaction with pyrite ash)	Reactive Blue 4	100%/0.5	[49]
UV radiation and ozonation	Reactive Blue 19	100%/0.1	[50]
AOP (Fenton, photo-Fenton), UV radiation	Reactive Blue 19	81%, 98%, 42%/0.3	[51]
AOP (Sulfate radical + UV)	Acid Blue 129	87%/1	This work

2. Results and Discussion

Several experiments were performed, including the influence of •OH and $SO_4^{•-}$ on the decolorization efficiency, effect of PDS concentration, pH, possible reaction pathways, and the ecotoxicity of by-products.

2.1. Influence of •OH and $SO_4^{•-}$ on AB129 Decolorization

To determine the decolorization efficiency of both radical species on AB129, we performed experiments with PDS (source of $SO_4^{•-}$) and H_2O_2 (•OH) catalyzed by UV. It is known that from 1 mole of oxidant, 2 moles of radicals may be generated, according to Equations (1) and (2):

$$S_2O_8^{2-} \xrightarrow{h\nu} 2SO_4^{•-} \tag{1}$$

$$H_2O_2 \xrightarrow{h\nu} 2HO^{•} \tag{2}$$

Figure 1 shows that UV irradiation alone [similarly to PDS (Figure 2) and H_2O_2 (data not shown) w/o UV activation] was not able to degrade the dye. The dye seems to be resistant to direct UV photolysis, as the energy of the photons with a wavelength ranging from 313 to 578 nm is too low to degrade the molecule of the dye. Also, decolorization by •OH is relatively slow and reached only 12% after one hour of the experiment. Only the UV-catalyzed $SO_4^{•-}$ oxidation process was found to be effective in the decolorization of AB129, whereby the effect was about 90% of the initial dye concentration (25 mg/L). Homolysis of the peroxide bond of PDS occurs when catalyzed by UV, which results in the generation of $SO_4^{•-}$ [52]. In proposed UV/PDS system (in near neutral or acidic pH) the only type of free radicals formed could be $SO_4^{•-}$. Water molecules could be oxidized to produce •OH but this process is very slow ($k = 6.6 \times 10^2$ s^{-1}) [53] and therefore, not significant in the timeframe of the experiment.

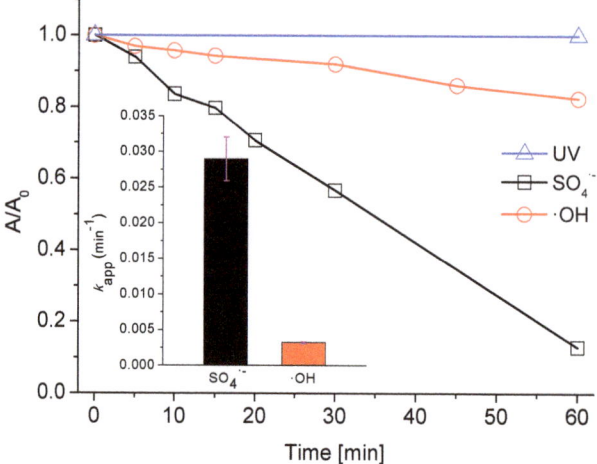

Figure 1. Decolorization (absorbance at 595 nm) of AB129 with $SO_4^{•-}$, •OH and ultraviolet (UV) radiation alone (conditions: 25 mg/L AB129, 2.5 mM PDS, 10 mM H_2O_2, UV 150 W). The inset shows decolorization kinetics of AB129 by $SO_4^{•-}$ and •OH, (the fuchsia error bars represent the slope error).

Despite the H_2O_2 concentrations being four times higher than in the case of PDS (10 mM vs. 2.5 mM), the generated •OH radicals did not react with the AB129 as effectively due to the following possible reasons. Although H_2O_2 and PDS molecules have a similar bond length of 1.453 Å and 1.497 Å [54], H_2O_2 peroxide bond energy (51.0 kcal/mol) is significantly higher than PDS (33.5 kcal/mol) and, therefore, it is more difficult to be cleaved by UV irradiation [55]. Furthermore, •OH has an almost ten times faster recombination rate ($k = 5.2 \times 10^9$ M^{-1} s^{-1}) [56] than $SO_4^{•-}$ ($k = 5.0 \times 10^8$ M^{-1} s^{-1}) [57] and, therefore, a smaller amount of generated radicals is available to react with the contaminant, compared to $SO_4^{•-}$. Further differences in the decolorization of AB129 may be due to intrinsic differences in the reaction mechanisms. While $SO_4^{•-}$ works more by electron abstraction because of a higher electron affinity (2.43 eV) than $^{\bullet}$OH (1.83 eV), $^{\bullet}$OH acts more through hydrogen abstraction or addition [58]. This makes $SO_4^{\bullet-}$ more selective and highly reactive towards organic contaminants containing non-bonding electron pairs of atoms such as O, N and S [59]. The first steps of the reaction between AB129 and sulfate radicals are described in more detail in the subsection "Formation of by-products".

The apparent first order (k_{app}) rate constants, shown in the inset of Figure 1 and calculated based on Equation (9), are one order of magnitude different higher for $SO_4^{\bullet-}$ than for $^{\bullet}$OH (0.029 min^{-1} and 0.0032 min^{-1} for $SO_4^{\bullet-}$ and $^{\bullet}$OH, respectively). Both radicals are susceptible to electron transfer; however, $SO_4^{\bullet-}$ shows a much lower energy barrier for this reaction, which results in markedly higher k_{app}. Therefore, it was decided to focus solely on PDS for a better understanding of its reaction mechanism with the AB129 dye.

2.2. Effect of PDS Concentration

To determine the optimal decolorization conditions, the concentration of PDS was changed from 0.625 to 2.5 mM, as depicted in Figure 2, where the inset shows the decolorization kinetics of AB129 by different PDS concentrations.

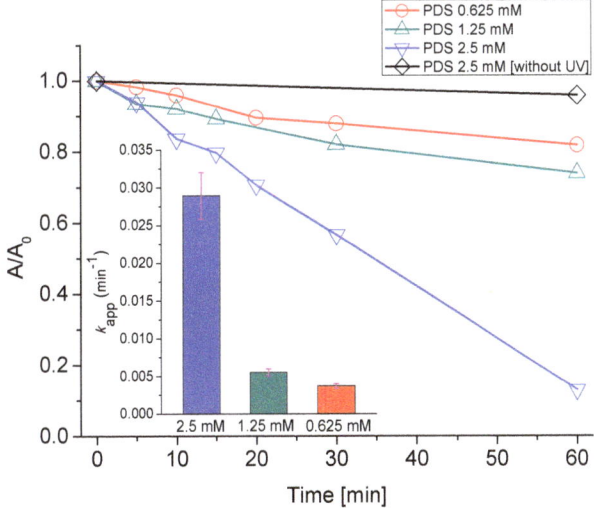

Figure 2. Decolorization (absorbance at 595 nm) of AB129 by UV/peroxydisulfate (PDS) system (25 mg/L AB129, UV 150 W). The fuchsia error bars represent the slope error

The concentrations of 0.625 mM and 1.25 mM achieved only 18% and 26% decolorization, respectively, and are almost comparable with the blank experiment w/o UV light. A further increase to 2.5 mM caused a significant improvement. The kinetic of the dye removal is significantly faster, the decrease is linear, and after 60 min the dye decolorization reached 87%. The apparent first-order rate constants calculated were 0.0037 min^{-1}, 0.0056 min^{-1} and 0.029 min^{-1} for 0.625 mM, 1.25 mM and 2.5 mM PDS, respectively. Therefore, for the four-fold increase in the PDS concentration, the rate constant increased roughly eight times. This may be because the low concentrations of oxidant did not produce enough $SO_4^{\bullet-}$ to degrade the dye effectively [60]. Finally, 2.5 mM was chosen as the optimal concentration in the experiment in terms of efficiency and economy, because higher PDS concentrations are not economically feasible.

2.3. Effect of the Initial pH

The other parameter that significantly influences the decolorization efficiency is the initial pH of the solution. The initial solution pH was varied in the interval between 3 and 11, as shown in Figure 3.

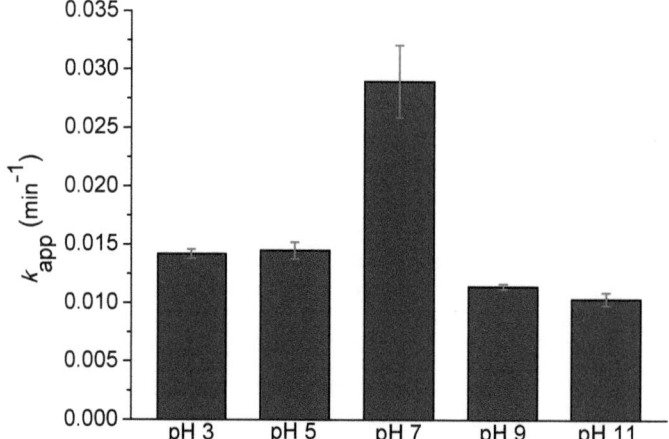

Figure 3. Effect of the initial pH on AB129 decolorization rate constant (conditions: 25 mg/L AB129, absorbance measured at 595 nm, UV 150 W). The fuchsia error bars represent the slope error.

It can be inferred that the pH conditions had a significant influence on the UV catalyzed PDS oxidation system. The apparent first-order rate constant increased noticeably from the pH range of 3 to 5 (0.0141 min^{-1} and 0.0145 min^{-1}) to pH 7 (0.029 min^{-1}) and decreased back to half-values for a higher pH (0.0114 min^{-1} and 0.0107 min^{-1} for pH 9 and 11, respectively). The results show that the neutral conditions are the most optimal for the decolorization reaction. Under an alkaline pH, the hydroxides (OH$^-$) in the solution undergo reactions with SO$_4^{\bullet-}$ to generate $^{\bullet}$OH (Equation (3)), which is a significantly less effective radical species in this respect.

$$SO_4^{\bullet-} + OH^- \rightarrow {}^{\bullet}OH + SO_4^{2-} \tag{3}$$

Furthermore, the generated $^{\bullet}$OH further reacts with SO$_4^{\bullet-}$ (Equation (4)), decreasing the number of available radicals.

$$SO_4^{\bullet-} + {}^{\bullet}OH \rightarrow HSO_5^- \tag{4}$$

Under an acidic pH, the further breakdown of PDS to SO$_4^{\bullet-}$ may be catalyzed by acid activation (Equation (5) and (6)).

$$S_2O_8^{2-} + H^+ \rightarrow HS_2O_8^- \tag{5}$$

$$HS_2O_8^- \rightarrow SO_4^{\bullet-} + SO_4^{2-} + H^+ \tag{6}$$

However, the generation of SO$_4^{\bullet-}$ catalyzed by acid conditions and UV together would yield high concentrations of those radicals. In excess, SO$_4^{\bullet-}$ may favor reactions like scavenging (Equation (7)) [61] or recombination (Equation (8)) over reactions with the dye.

$$S_2O_8^{2-} + SO_4^{\bullet-} \rightarrow S_2O_8^{\bullet-} + SO_4^{2-} \tag{7}$$

$$SO_4^{\bullet-} + SO_4^{\bullet-} \rightarrow S_2O_8^{2-} \tag{8}$$

This may explain k_{app} being lower in an acidic pH compared to a neutral pH, which was also observed by Liang et al. [62] in a PDS oxidation system. Overall, the most favorable condition for oxidation of AB129 is at pH of 7. Alkaline and acidic pH conditions caused inhibition of the reaction by the possible reasons explained. After evaluating the effect of pH on the decolorization process, the study focused on the identification of post-treatment intermediates.

2.4. Formation of by-Products

To determine the possible formation of by-products, the absorbance spectra during the decolorization of AB129 were recorded, as depicted in Figure 4.

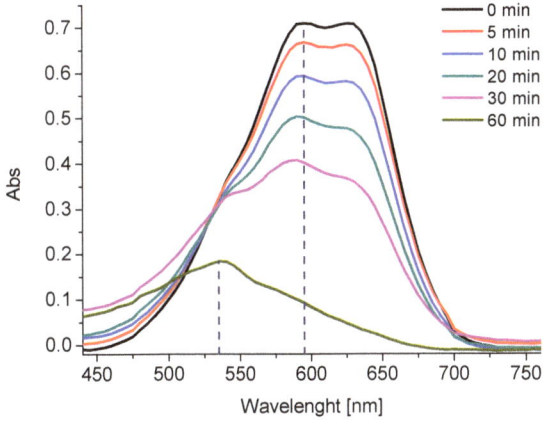

Figure 4. Absorbance spectra of AB129 during decolorization tests (conditions: 25 mg/L AB129, 2.5 mM PDS).

At the beginning, a double peak at 595 and 630 nm was recorded. During the experiment, the peak at 630 nm slowly disappeared, followed by the peak at 595 nm. At 20 min, a new peak at 535 nm was formed, which dominated at the end of the experiment (60 min). This peak may represent the formed by-products.

Quantum chemical calculations were performed to determine the most probable pathway of the reaction between the sulfate radical and AB129. Firstly, the geometry of AB129 was optimized, as shown in Figure 5.

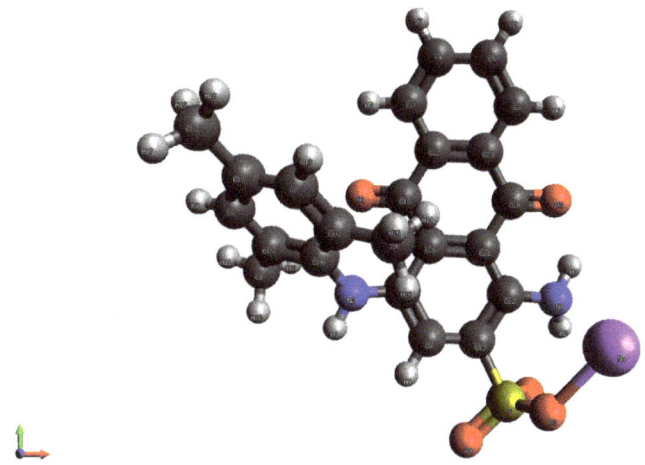

Figure 5. Optimized AB129 molecule obtained using B3LYP/6-31G**.

Then the CPCM approach was applied in order to model the solvent (water) effect on the calculated transition energies of the species with the def2-TZVP basis set. The λ_{max} of the visible spectrum was computed to be 594 nm, almost the same as the wavelength, which was used for determination of AB129 (595 nm), and corresponded to the HOMO → LUMO and HOMO-1 → LUMO (overall E = 2.087 eV) transitions of AB129. Moreover, the values of HOMO and LUMO were found to be −5.413 and −2.863 eV, respectively, whereas the difference in the energy (HOMO-LUMO energy gap) was 2.55 eV. Similar values were recently computed and reported for Acid Blue 113 by Asghar et al. [63]. Figure 6 shows the HOMO and LUMO of AB129 obtained at B3LYP/6-31G** and a map of the electron density of the AB129 molecule.

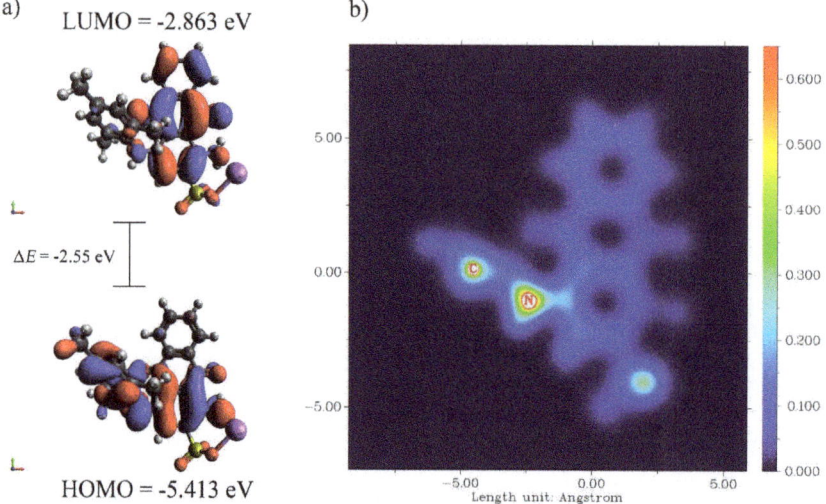

Figure 6. (a) HOMO and LUMO of AB129 obtained at B3LYP/6-31G** and (b) map of the electron density (X, Y plane) of AB129 molecule (the sodium atom was removed for simplicity).

According to the frontier orbital theory, chemical reactions preferentially occur at the position of the molecule wherein their frontier orbital intensely overlap [64]. Moreover, the most probable reaction pathway for sulfate radicals that have a very strong electrophilic character is a direct attack on one of the atoms of the contaminant molecule, usually the one with the highest electron density in the HOMO of the aromatic molecule [65]. Figure 6b shows that one of the positions with the highest electron densities (in the HOMO of AB129) is the region near to nitrogen atom from the secondary amine. From this, it is possible to conclude that there is a higher preference for the –NH- group, and that the main product forming in this system is a derivative of hydroxylated anthraquinone.

Furthermore, according to Liu et al. [66], a Hirshfeld charge may be successfully employed to determine the reactive sites of the electrophilic reactions. Apart from the oxygen atoms, which are probably not involved in the reactions reported in this study, and a high O/C ratio is often correlated with a slow reaction between the molecules and the oxidants [67], the N atom of AB129 may be characterized by the smallest Hirshfeld charge (−0.424), which is even smaller than the second nitrogen (N1: −0.398) from the primary amine located on the anthraquinone. This may provide further confirmation that the first and crucial reaction of the sulfate radical with AB129 is an electron transfer from the –NH- moiety, which splits the AB129 molecule and creates the anthraquinone derivative.

This conclusion may be supported in several other ways. Primarily, the intermediate that was formed absorbs photons of higher energy (Vis peak shifted to the left side of the spectrum i.e., 535 nm), which is typical for the anthraquinone derivatives with a much lower molecular weight [39]. A similar observation was made by Tang and An, who observed radical driven splitting of Acid Blue 40 with the formation of a yellow intermediate absorbing light in a similar region to that reported in this study [39].

Considering the possible formation of by-products, their toxicity on model plant fronds and freshwater crustaceans was evaluated and is discussed below.

2.5. Ecotoxicity

In addition to a determination of decolorization efficiency, it is important to consider the toxicity of the system for living organisms, since post-treatment by-products may sometimes be more toxic than the initial contaminant [41]. Tests using plant fronds *Daphnia magna* and freshwater crustaceans *Lemna minor* are often performed in toxicological studies, because they are simple, fast and cost-effective. Moreover, they represent both plants and animals, which may tell us more about the impact on the ecosystem, and selected microorganisms are very informative in terms of the potential toxicity of wastewater [68–70]. For example, Sackey et al. found that *Daphnia magna* and *Lemna minor* are effective for testing the toxicity of leachates [71]. Moreover, Castro et al. [72] investigated the potential toxicity of effluents from the textile industry before and after treatment, and concluded that the raw textile effluent was very toxic. Therefore, toxicity tests were performed using the same conditions mentioned in the methodology of the decolorization test, except different concentrations of PDS and AB129 were used, as shown in Tables 3 and 4.

Table 3. Toxicity of AB129 by-products on *Daphnia magna* (* Time 0 min = samples without PDS addition).

	Daphnia Magna					
	Time [min]					
PDS/AB129 [mM]	0 *	5	10	20	30	45
	Toxicity Effect [%]					
0.2/0.2	25	40	35	30	25	20
0.5/0.5	30	40	45	40	30	20
1/1	40	55	65	50	35	25
2/2	50	60	65	55	40	30

Table 4. Toxicity of AB129 by-products on *Lemna minor* (* Time 0 min = samples without PDS addition).

	Lemna Minor					
	Time [min]					
PDS/AB129 [mM]	0 *	5	10	20	30	45
	Toxicity Effect [%]					
0.2/0.2	8	17	8	0	0	0
0.5/0.5	25	25	25	8	0	0
1/1	33	25	33	17	8	8
2/2	33	33	25	17	17	17

According the guidelines for the interpretation of the obtained toxicity results given by Kudlek [73], samples characterized by a toxic effect of <25% are nontoxic. Only the lowest concentration of PDS/AB129 (0.2/0.2 mM) was nontoxic for the *Lemna minor* test organisms, whereas *Daphnia magna* organisms were more sensitive to the action of PDS/AB129, and classified the post-processed samples subjected to both 0.2/0.2 mM and 0.5/0.5 mM as low toxic (a toxicity effect of between 25% and 50%). A toxicity effect higher than 50% classified the samples as being toxic. Such results were noted in the samples between 5 and 20 min of the experiment tested on *Daphnia magna*, where the concentration of PDS/AB129 was equal to 1/1 mM and 2/2 mM.

Nonetheless, both tests indicated that the toxicity of the initial solutions increased along with the PDS/AB129 concentration. In both tests, up to approximately 10 min of the experiment, the toxicity increased due to the effect of the addition of PDS. However, in both cases, after this time, a decreasing trend in toxicity was detected. After 45 min of the experiment, the toxicity for *Daphnia* roughly halved for all of the concentrations analyzed and for *Lemna* it decreased even more. This may indicate that by-products of AB129 after treatment are less toxic than the original dye. Moreover, PDS toxicity is only temporary, because it is quickly decomposed and exhibits lower toxicity to the analyzed organisms.

3. Materials and Methods

3.1. Chemicals

Sodium persulfate ($Na_2S_2O_8$, purity ≥98%), hydrogen peroxide (H_2O_2, 30% w/w in water), sodium hydroxide (NaOH, 97% powder), sulfuric acid (H_2SO_4, 95%–98%), and Acid blue (AB129, $C_{23}H_{19}N_2NaO_5S$, 25% dye content) were purchased from Sigma Aldrich (Prague, Czech Republic). Hydrochloric acid (HCl, >35%) was purchased from Avantor Performance Materials Poland (Gliwice, Poland). Deionized water (18.2 MΩ·cm) obtained from ELGA purelab flex system (ELGA, Veolia Water, Marlow, UK) was used in all of the experiments.

3.2. Analytical

A pH meter TMultiLine® Multi 3430 IDS from WTW (Weilheim, Germany) equipped with SenTix pH electrodes was used to measure the acidity of the reaction mixture. The visible spectrum of the samples was measured by a UV-Vis spectrophotometer DR 3900 from Hach (Vancouver, WA, USA) within the 440–760 nm wavelength range, recorded every 5 nm.

3.3. Decolorization Test

Decolorization experiments of AB129 were performed based on a modified method of Neamtu et al. [74]. Firstly, a solution of AB129 (25 mg/L) and PDS (various concentrations of 0.625 mM, 1.25 mM and 2.5 mM) or H_2O_2 (10 mM) was prepared in a 100 mL reactor. Then, pH conditions were adjusted by adding a minimal amount of concentrated NaOH or H_2SO_4 solution, and the prepared reactor was exposed to UV radiation under constant magnetic stirring. The UV light source was provided by a model TQ 150 medium-pressure mercury UV lamp (Heraeus, Hanau, Germany) placed in a quartz glass (DURAN 50) cooling jacket fed by recirculating tap water. This step maintained a constant temperature of the mixture of 21 ± 1 °C. According to the data provided by the manufacturer, the TQ 150 lamp operated in the cooling jacket emanates radiation with a wavelength λ_{exc} of 313, 365, 405, 436, 546, 578 nm, and radiation flux equal to 2.5, 5.8, 2.9, 3.6, 4.6, 4.2 W, respectively. The absorbance spectra were measured in 1 mL quartz cuvettes by a UV-Vis spectrometer at the wavelength of 595 nm according to Palencia et al. [75]. The analyses were performed several times and averages and standard deviations were calculated by Origin 9 software [OriginLab].

3.4. Kinetic Test and AB129 Structure Modelling

A pseudo-first order kinetic model was used to describe the decolorization of AB129 by $SO_4^{\bullet -}$ and $^\bullet OH$ (Equation (9)) [76].

$$\ln\left(\frac{C_t}{C_0}\right) = \ln\left(\frac{A_t}{A_0}\right) = -k_{app} t \qquad (9)$$

where C_0 and C_t are the initial ($t = 0$) and time-dependent concentrations (at time t), proportional to the measured absorbance A, respectively, and k_{app} is an apparent rate constant [77].

3.5. Quantum Chemical Analysis

The initial coordinates of the AB129 structure were obtained with the Avogadro program [78]. The structure of the AB129 was further optimized using the Orca program package [79], and the results were validated with Gaussian 16 software [63], both in the gaseous and liquid phase at the B3LYP/6-31G** level of study, as suggested in a recent work on the Acid blue 113 oxidation [63]. Time-dependent density functional theory TD-DFT was used to predict the excited state properties of AB129. The outputs were later visualized with the Avogadro program, whereas the electron densities and Hirshfeld charges were visualized and computed by Multiwfn software [80,81].

3.6. Ecotoxicity Test

Two different bio-tests were used to determine the toxicity of the post-treatment products: the *Lemna* sp. growth inhibition test (GIT) and the Daphtoxkit F bioassay. In the GIT, plant fronds of freshwater vascular plants *Lemna minor* from our own breeding were used. The test is based on calculating the number of plant fronds growing for 7 days in a tested and blank sample, prepared according to the OECD Guideline 221. The test was performed at a temperature of 25 ± 1 °C by a constant exposure to light with an illuminance of 6000 lux.

The Daphtoxkit F bioassay from Tigret (Warszawa, Poland) uses freshwater crustaceans *Daphnia magna* to measure their immobility or mortality after 24 h exposure to tested post-process samples, in comparison to standard freshwater (ISO medium prepared according ISO 6341). The test was performed on 1-day-old test organisms, according to the OECD Guideline 202. NaOH (0.1 mol/L) and HCl (0.1 mol/L) solutions were used for pH corrections during the toxicity tests. The toxicity of both tests was calculated using the following equation 10 [73]:

$$E = \frac{(N_C - N_T)}{N_C} \cdot 100\% \qquad (10)$$

where E is the toxicity effect (%), N_C is the number of living organisms (plant fronds or freshwater crustaceans) in the control sample, and N_T is the number of living organisms (plant fronds or freshwater crustaceans) in the test sample. Interpretation of the results obtained from both of the toxicity tests was performed based on the toxicity classification presented in Table 5, and according to guidelines proposed by Mahugo Santana et al. [82].

Table 5. Interpretation of the toxicity results.

Toxicity Effect E%	Toxicity Classification
<25	non-toxic
25–50	low toxic
50.01–75	toxic
>75	highly toxic

4. Conclusions

In this work, we focused on sulfate and hydroxyl radical-based oxidation processes catalyzed by UV for the treatment of the model dye Acid Blue 129. $SO_4^{\bullet-}$ at a concentration of 2.5 mM successfully decolorized 25 mg/L of the dye up to 87% within 60 min, whereas $^{\bullet}OH$ at a concentration of 10 mM was significantly less effective. The pseudo-first-order kinetic rate constant of the optimal reaction conditions, including neutral pH, was found to be 0.029 min^{-1}. The probable reaction pathway of AB129 with $SO_4^{\bullet-}$ was determined using quantum chemical calculations, indicating electron transfer from the –NH- moiety, which splits the AB129 molecule creating the anthraquinone derivative. Ecotoxicity tests of the by-products showed a lower toxicity than the toxicity of the initial dye and only a temporary effect of PDS.

Author Contributions: Conceptualization, S.W. and K.K.; methodology, K.K. and S.W.; software, S.W.; validation, S.W., E.K. and M.Č.; investigation, T.K. and K.K.; writing—original draft preparation, K.K. and E.K.; writing—review and editing, K.G., S.W., D.S., V.V.T.P., and M.Č.; supervision, S.W. and M.Č. All authors have read and agreed to the published version of the manuscript

Funding: This research was supported by the Ministry of Education, Youth and Sports in the Czech Republic under the "Inter Excellence – Action programme" within the framework of the project "Exploring the role of ferrates and modified nano zero-valent iron in the activation process of persulfates" (registration number LTAUSA18078) and the Research Infrastructures NanoEnviCz (Project No. LM2018124). This work was also supported by the Ministry of Education, Youth and Sports of the Czech Republic and the European Union - European Structural and Investment Funds in the frames of Operational Programme Research, Development and Education - project Hybrid Materials for Hierarchical Structures (HyHi, Reg. No. CZ.02.1.01/0.0/0.0/16_019/0000843). Finally, the work was supported by the Ministry of Science and Higher Education Republic of Poland within statutory funds No. 08/040/BK_19/0119, and by the National Centre for Research and Development in Poland (POIR.04.01.02-00-0062/16).

Conflicts of Interest: The authors declare no conflict of interest.

References

1. Pal, A.; He, Y.; Jekel, M.; Reinhard, M.; Gin, K.Y.H. Emerging contaminants of public health significance as water quality indicator compounds in the urban water cycle. *Environ. Int.* **2014**, *71*, 46–62. [CrossRef]
2. Tsiampalis, A.; Frontistis, Z.; Binas, V.; Kiriakidis, G.; Mantzavinos, D. Degradation of sulfamethoxazole using iron-doped titania and simulated solar radiation. *Catalysts* **2019**, *9*, 612. [CrossRef]
3. Şen, S.; Demirer, G.N. Anaerobic treatment of real textile wastewater with a fluidized bed reactor. *Water Res.* **2003**, *37*, 1868–1878. [CrossRef]
4. Ozturk, E.; Yetis, U.; Dilek, F.B.; Demirer, G.N. A chemical substitution study for a wet processing textile mill in Turkey. *J. Clean. Prod.* **2009**, *17*, 239–247. [CrossRef]
5. Inoue, M.; Okada, F.; Sakurai, A.; Sakakibara, M. A new development of dyestuffs degradation system using ultrasound. *Ultrason. Sonochem.* **2006**, *13*, 313–320. [CrossRef]
6. Eren, Z. Ultrasound as a basic and auxiliary process for dye remediation: A review. *J. Environ. Manag.* **2012**, *104*, 127–141. [CrossRef]
7. Julkapli, N.M.; Bagheri, S.; Hamid, S.B.A. Recent advances in heterogeneous photocatalytic decolorization of synthetic dyes. *Sci. World J.* **2014**, *2014*, 692307. [CrossRef]
8. Ghaly, A.E.; Ananthashankar, R.; Alhattab, M.; Ramakrishnan, V.V. Production, Characterization and Treatment of Textile Effluents: A Critical Review. *J. Chem. Eng. Process. Technol.* **2014**, *5*, 1000182.
9. Chequer, F.D.; de Oliveira, G.A.R.; Ferraz, E.R.A.; Carvalho, J.; Zanoni, M.B.; de Oliveir, D.P. Textile Dyes: Dyeing Process and Environmental Impact. *Eco-Friendly Text. Dye. Finish.* **2013**, *6*, 151–176.
10. Wang, J.; Guo, B.; Zhang, X.; Zhang, Z.; Han, J.; Wu, J. Sonocatalytic degradation of methyl orange in the presence of TiO_2 catalysts and catalytic activity comparison of rutile and anatase. *Ultrason. Sonochem.* **2005**, *12*, 331–337. [CrossRef]
11. Wijetunga, S.; Li, X.F.; Jian, C. Effect of organic load on decolourization of textile wastewater containing acid dyes in upflow anaerobic sludge blanket reactor. *J. Hazard. Mater.* **2010**, *177*, 792–798. [CrossRef]
12. Entezari, M.H.; Al-Hoseini, Z.S.; Ashraf, N. Fast and efficient removal of Reactive Black 5 from aqueous solution by a combined method of ultrasound and sorption process. *Ultrason. Sonochem.* **2008**, *15*, 433–437. [CrossRef] [PubMed]
13. Crini, G.; Badot, P.-M. *Sorption Processes and Pollution: Conventional and Non-Conventional Sorbents for Pollutant Removal from Wastewaters*; Presses universitaires de Franche-Comté: Besançon, France, 2010; ISBN 2848673044.
14. Puvaneswari, N.; Muthukrishnan, J.; Gunasekaran, P. Toxicity assessment and microbial degradation of azo dyes. *Indian J. Exp. Biol.* **2006**, *44*, 618–626.
15. Oh, S.W.; Kang, M.N.; Cho, C.W.; Lee, M.W. Detection of carcinogenic amines from dyestuffs or dyed substrates. *Dye. Pigment.* **1997**, *33*, 119–135. [CrossRef]
16. Fat'hi, M.R.; Asfaram, A.; Hadipour, A.; Roosta, M. Kinetics and thermodynamic studies for removal of Acid Blue 129 from aqueous solution by almond shell. *J. Environ. Health Sci. Eng.* **2014**, *12*, 26. [CrossRef] [PubMed]
17. Acid Blue 129. Available online: http://datasheets.scbt.com/sc-214468.pdf (accessed on 22 April 2020).
18. Pala, A.; Tokat, E. Color removal from cotton textile industry wastewater in an activated sludge system with various additives. *Water Res.* **2002**, *36*, 2920–2925. [CrossRef]

19. Harrelkas, F.; Azizi, A.; Yaacoubi, A.; Benhammou, A.; Pons, M.N. Treatment of textile dye effluents using coagulation-flocculation coupled with membrane processes or adsorption on powdered activated carbon. *Desalination* **2009**, *235*, 330–339. [CrossRef]
20. Cooper, P. Removing colour from dyehouse waste waters—A critical review of technology available. *J. Soc. Dye. Colour.* **1993**, *109*, 97–100. [CrossRef]
21. Fane, A.G.; Fell, C.J.D. A review of fouling and fouling control in ultrafiltration. *Desalination* **1987**, *62*, 117–136. [CrossRef]
22. Georgiou, D.; Melidis, P.; Aivasidis, A. Use of a microbial sensor: Inhibition effect of azo-reactive dyes on activated sludge. *Bioprocess Biosyst. Eng.* **2002**, *25*, 79–83.
23. Katheresan, V.; Kansedo, J.; Lau, S.Y. Efficiency of various recent wastewater dye removal methods: A review. *J. Environ. Chem. Eng.* **2018**, *6*, 4676–4697. [CrossRef]
24. Li, Q.; Wang, L.; Fang, X.; Zhang, L.; Li, J.; Xie, H. Synergistic effect of photocatalytic degradation of hexabromocyclododecane in water by UV/TiO$_2$/persulfate. *Catalysts* **2019**, *9*, 189. [CrossRef]
25. Tehrani, A.R.; Mahmood, N.M.; Arami, M. Study of the efficiency of effective parameters on decolorization of CI. Reactive black 5 wastewater by ozonation. *J. Color Sci. Technol.* **2008**, *2*, 67–75.
26. Muruganandham, M.; Swaminathan, M. Photochemical oxidation of reactive azo dye with UV-H$_2$O$_2$ process. *Dye. Pigment.* **2004**, *62*, 269–275. [CrossRef]
27. Mohey El-Dein, A.; Libra, J.A.; Wiesmann, U. Mechanism and kinetic model for the decolorization of the azo dye Reactive Black 5 by hydrogen peroxide and UV radiation. *Chemosphere* **2003**, *52*, 1069–1077. [CrossRef]
28. Lovato, M.E.; Gilliard, M.B.; Cassano, A.E.; Martín, A.M. Kinetics of the degradation of n-butyl benzyl phthalate using O$_3$/UV, direct photolysis, direct ozonation and UV effects. *Environ. Sci. Pollut. Res.* **2015**, *22*, 909–917. [CrossRef]
29. Tehrani-Bagha, A.R.; Amini, F.L. Decolorization of a Reactive Dye by UV-Enhanced Ozonation. *Prog. Col. Color. Coat.* **2010**, *3*, 1–8.
30. Ghernaout, D. Advanced oxidation phenomena in electrocoagulation process: A myth or a reality? *Desalin. Water Treat.* **2013**, *51*, 7536–7554. [CrossRef]
31. Argun, M.E.; Karatas, M. Application of Fenton process for decolorization of reactive black 5 from synthetic wastewater: Kinetics and thermodynamics. *Environ. Prog. Sustain. Energy* **2011**, *30*, 540–548. [CrossRef]
32. Meriç, S.; Kaptan, D.; Ölmez, T. Color and COD removal from wastewater containing Reactive Black 5 using Fenton's oxidation process. *Chemosphere* **2004**, *54*, 435–441. [CrossRef]
33. Li, J.; Li, R.; Zou, L.; Liu, X. Efficient degradation of norfloxacin and simultaneous electricity generation in a persulfate-photocatalytic fuel cell system. *Catalysts* **2019**, *9*, 835. [CrossRef]
34. Rickman, K.A.; Mezyk, S.P. Kinetics and mechanisms of sulfate radical oxidation of β-lactam antibiotics in water. *Chemosphere* **2010**, *81*, 359–365. [CrossRef] [PubMed]
35. Olmez-Hanci, T.; Arslan-Alaton, I. Comparison of sulfate and hydroxyl radical based advanced oxidation of phenol. *Chem. Eng. J.* **2013**, *224*, 10–16. [CrossRef]
36. Zheng, J.; Li, J.; Bai, J.; Xiaohantan, X.; Zeng, Q.; Li, L.; Zhou, B. Efficient degradation of refractory organics using sulfate radicals generated directly from WO$_3$ photoelectrode and the catalytic reaction of sulfate. *Catalysts* **2017**, *7*, 346. [CrossRef]
37. Neta, P.; Madhavan, V.; Zemel, H.; Fessenden, R.W. Rate Constants and Mechanism of Reaction of SO$_4$ with Aromatic Compounds. *J. Am. Chem. Soc.* **1977**, *99*, 163–164. [CrossRef]
38. Buxton, G.V.; Greenstock, C.L.; Helman, W.P.; Ross, A.B. Critical Review of rate constants for reactions of hydrated electrons, hydrogen atoms and hydroxyl radicals (·OH/·O$^-$ in Aqueous Solution. *J. Phys. Chem. Ref. Data* **1988**, *17*, 513. [CrossRef]
39. Tang, W.Z. Huren an Photocatalytic degradation kinetics and mechanism of acid blue 40 by TiO$_2$/UV in aqueous solution. *Chemosphere* **1995**, *31*, 4171–4183. [CrossRef]
40. Li, X.; Tang, S.; Yuan, D.; Tang, J.; Zhang, C.; Li, N.; Rao, Y. Improved degradation of anthraquinone dye by electrochemical activation of PDS. *Ecotoxicol. Environ. Saf.* **2019**, *177*, 77–85. [CrossRef]
41. Wacławek, S.; Lutze, H.V.; Grübel, K.; Padil, V.V.T.; Černík, M.; Dionysiou, D.D. Chemistry of persulfates in water and wastewater treatment: A review. *Chem. Eng. J.* **2017**, *330*, 44–62. [CrossRef]
42. Yang, J.; Zeng, Z.; Huang, Z.; Cui, Y. Acceleration of persulfate activation by MIL-101(Fe) with vacuum thermal activation: Effect of FeII/FeIII mixed-valence center. *Catalysts* **2019**, *9*, 906. [CrossRef]

43. Duan, X.; Sun, H.; Kang, J.; Wang, Y.; Indrawirawan, S.; Wang, S. Insights into heterogeneous catalysis of persulfate activation on dimensional-structured nanocarbons. *ACS Catal.* **2015**, *5*, 4629–4636. [CrossRef]
44. Ji, Y.; Shi, Y.; Yang, Y.; Yang, P.; Wang, L.; Lu, J.; Li, J.; Zhou, L.; Ferronato, C.; Chovelon, J.M. Rethinking sulfate radical-based oxidation of nitrophenols: Formation of toxic polynitrophenols, nitrated biphenyls and diphenyl ethers. *J. Hazard. Mater.* **2019**, *361*, 152–161. [CrossRef] [PubMed]
45. Gözmen, B.; Kayan, B.; Gizir, A.M.; Hesenov, A. Oxidative degradations of reactive blue 4 dye by different advanced oxidation methods. *J. Hazard. Mater.* **2009**, *168*, 129–136. [CrossRef] [PubMed]
46. Lizama, C.; Freer, J.; Baeza, J.; Mansilla, H.D. Optimized photodegradation of reactive blue 19 on TiO_2 and ZnO suspensions. *Catal. Today* **2002**, *76*, 235–246. [CrossRef]
47. Tehrani-Bagha, A.R.; Mahmoodi, N.M.; Menger, F.M. Degradation of a persistent organic dye from colored textile wastewater by ozonation. *Desalination* **2010**, *260*, 34–38. [CrossRef]
48. Bilal, M.; Rasheed, T.; Iqbal, H.M.N.; Li, C.; Wang, H.; Hu, H.; Wang, W.; Zhang, X. Photocatalytic degradation, toxicological assessment and degradation pathway of C.I. Reactive Blue 19 dye. *Chem. Eng. Res. Des.* **2018**, *129*, 384–390. [CrossRef]
49. Becelic-Tomin, M.; Dalmacija, B.; Rajic, L.; Tomasevic, D.; Kerkez, D.; Watson, M.; Prica, M. Degradation of anthraquinone dye reactive blue 4 in pyrite ash catalyzed fenton reaction. *Sci. World J.* **2014**, *2014*, 234654. [CrossRef]
50. Lovato, M.E.; Fiasconaro, M.L.; Martín, C.A. Degradation and toxicity depletion of RB19 anthraquinone dye in water by ozone-based technologies. *Water Sci. Technol.* **2017**, *75*, 813–822. [CrossRef]
51. Radovic, M.; Mitrovic, J.; Kostic, M.; Bojic, D.; Petrovic, M.; Najdanovic, S.; Bojic, A. Comparison of ultraviolet radiation/hydrogen peroxide, Fenton and photo-Fenton processes for the decolorization of reactive dyes. *Hem. Ind.* **2015**, *69*, 657–665. [CrossRef]
52. Sharma, S.; Patel, S.; Ruparelia, J. Feasibility study on degradation of RR120 dye from water by O_3, O_3/UV and O_3/UV/Persulfate. In *In Multi-Disciplinary Sustainable Engineering: Current and Future Trends, Proceedings of the 5th Nirma University International Conference on Engineering, Ahmedabad, India, 26–28 November 2015*; CRC Press: Boca Raton, FL, USA, 2016; p. 233.
53. Herrmann, H.; Reese, A.; Zellner, R. Time-resolved UV/VIS diode array absorption spectroscopy of SO_x^- (x=3, 4, 5) radical anions in aqueous solution. *J. Mol. Struct.* **1995**, *348*, 183–186. [CrossRef]
54. Tan, C.; Gao, N.; Deng, Y.; Zhang, Y.; Sui, M.; Deng, J.; Zhou, S. Degradation of antipyrine by UV, UV/H_2O_2 and UV/PS. *J. Hazard. Mater.* **2013**, *260*, 1008–1016. [CrossRef] [PubMed]
55. Shah, N.S.; He, X.; Khan, H.M.; Khan, J.A.; O'Shea, K.E.; Boccelli, D.L.; Dionysiou, D.D. Efficient removal of endosulfan from aqueous solution by UV-C/peroxides: A comparative study. *J. Hazard. Mater.* **2013**, *263*, 584–592. [CrossRef] [PubMed]
56. Neta, P.; Huie, R.E.; Ross, A.B. Rate Constants for Reactions of Inorganic Radicals in Aqueous Solution. *J. Phys. Chem. Ref. Data* **1988**, *17*, 1027–1284. [CrossRef]
57. De Laat, J.; Le, T.G. Kinetics and modeling of the Fe(III)/H_2O_2 system in the presence of sulfate in acidic aqueous solutions. *Environ. Sci. Technol.* **2005**, *39*, 1811–1818. [CrossRef]
58. Yang, Z.; Su, R.; Luo, S.; Spinney, R.; Cai, M.; Xiao, R.; Wei, Z. Comparison of the reactivity of ibuprofen with sulfate and hydroxyl radicals: An experimental and theoretical study. *Sci. Total Environ.* **2017**, *590*, 751–760. [CrossRef]
59. Ghauch, A.; Baalbaki, A.; Amasha, M.; El Asmar, R.; Tantawi, O. Contribution of persulfate in UV-254 nm activated systems for complete degradation of chloramphenicol antibiotic in water. *Chem. Eng. J.* **2017**, *317*, 1012–1025. [CrossRef]
60. Silveira, J.E.; Garcia-Costa, A.L.; Cardoso, T.O.; Zazo, J.A.; Casas, J.A. Indirect decolorization of azo dye Disperse Blue 3 by electro-activated persulfate. *Electrochim. Acta* **2017**, *258*, 927–932. [CrossRef]
61. Lin, H.; Zhang, H.; Hou, L. Degradation of C. I. Acid Orange 7 in aqueous solution by a novel electro/Fe_3O_4/PDS process. *J. Hazard. Mater.* **2014**, *276*, 182–191. [CrossRef]
62. Liang, C.; Wang, Z.S.; Bruell, C.J. Influence of pH on persulfate oxidation of TCE at ambient temperatures. *Chemosphere* **2007**, *66*, 106–113. [CrossRef]
63. Asghar, A.; Bello, M.M.; Raman, A.A.A.; Daud, W.M.A.W.; Ramalingam, A.; Zain, S.B.M. Predicting the degradation potential of Acid blue 113 by different oxidants using quantum chemical analysis. *Heliyon* **2019**, *5*, e02396. [CrossRef]

64. Lang, A.R. *Dyes and Pigments: New Research*; Nova Science Publishers: Hauppauge, NY, USA, 2009; ISBN 978-1-60876-195-1.
65. Cinar, Z. The Role of Molecular Modeling in TiO_2 Photocatalysis. *Molecules* **2017**, *22*, 556. [CrossRef]
66. Liu, S.; Rong, C.; Lu, T. Information conservation principle determines electrophilicity, nucleophilicity, and regioselectivity. *J. Phys. Chem. A* **2014**, *118*, 3698–3704. [CrossRef] [PubMed]
67. Xiao, R.; Ye, T.; Wei, Z.; Luo, S.; Yang, Z.; Spinney, R. Quantitative Structure–Activity Relationship (QSAR) for the Oxidation of Trace Organic Contaminants by Sulfate Radical. *Environ. Sci. Technol.* **2015**, *49*, 13394–13402. [CrossRef] [PubMed]
68. Ziegler, P.; Sree, K.S.; Appenroth, K.J. Duckweeds for water remediation and toxicity testing. *Toxicol. Environ. Chem.* **2016**, *98*, 1127–1154. [CrossRef]
69. Žaltauskaitė, J.; Sujetovienė, G.; Čypaitė, A.; Aužbikavičiūtė, A. Lemna Minor as a Tool for Wastewater Toxicity Assessment and Pollutants Removal Agent. In Proceedings of the 9th International Conference on Environmental Engineering, Vilnius, Lithuania, 22–23 May 2014.
70. Mkandawire, M.; Teixeira Da Silva, J.A.; Dudel, E.G. The lemna bioassay: Contemporary issues as the most standardized plant bioassay for aquatic ecotoxicology. *Crit. Rev. Environ. Sci. Technol.* **2014**, *44*, 154–197. [CrossRef]
71. Sackey, L.N.A.; Kočí, V.; van Gestel, C.A.M. Ecotoxicological effects on Lemna minor and Daphnia magna of leachates from differently aged landfills of Ghana. *Sci. Total Environ.* **2020**, *698*, 134295. [CrossRef]
72. Castro, A.M.; Nogueira, V.; Lopes, I.; Rocha-Santos, T.; Pereira, R. Evaluation of the potential toxicity of effluents from the textile industry before and after treatment. *Appl. Sci.* **2019**, *9*, 3804. [CrossRef]
73. Kudlek, E. Identification of Degradation By-Products of Selected Pesticides during Oxidation and Chlorination Processes. *Ecol. Chem. Eng. S* **2019**, *26*, 571–581. [CrossRef]
74. Neamtu, M.; Siminiceanu, I.; Yediler, A.; Kettrup, A. Kinetics of decolorization and mineralization of reactive azo dyes in aqueous solution by the UV/H_2O_2 oxidation. *Dye. Pigment.* **2002**, *53*, 93–99. [CrossRef]
75. Palencia, M.; Martínez, J.M.; Arrieta, Á. Removal of Acid Blue 129 dye by Polymer-Enhanced Ultrafiltration (PEUF). *J. Sci. Technol. Appl.* **2017**, *2*, 65–74. [CrossRef]
76. Stumm, W.; Morgan, J.J. *Aquatic Chemistry: Chemical Equilibria and Rates in Natural Waters*, 3rd ed.; Wiley: Hoboken, NJ, USA, 2012; ISBN 1118591488.
77. Silvestri, D.; Wacławek, S.; Venkateshaiah, A.; Krawczyk, K.; Sobel, B.; Padil, V.V.T.; Černík, M.; Varma, R.S. Synthesis of Ag nanoparticles by a chitosan-poly(3-hydroxybutyrate) polymer conjugate and their superb catalytic activity. *Carbohydr. Polym.* **2020**, *232*, 115806. [CrossRef]
78. Hanwell, M.D.; Curtis, D.E.; Lonie, D.C.; Vandermeersch, T.; Zurek, E.; Hutchison, G.R. Avogadro: An advanced semantic chemical editor, visualization, and analysis platform. *J. Cheminform.* **2012**, *4*, 17. [CrossRef]
79. Neese, F. The ORCA program system. *WIRES Comput. Mol. Sci.* **2012**, *2*, 73–78. [CrossRef]
80. Lu, T.; Chen, F. Multiwfn: A multifunctional wavefunction analyzer. *J. Comput. Chem.* **2012**, *33*, 580–592. [CrossRef]
81. Fu, R.; Lu, T.; Chen, F.W. Comparing methods for predicting the reactive site of electrophilic substitution. *Acta Phys.-Chim. Sin.* **2014**, *30*, 628–639.
82. Santana, C.M.; Ferrera, Z.S.; Padrón, M.E.T.; Rodríguez, J.J.S. Methodologies for the extraction of phenolic compounds from environmental samples: New approaches. *Molecules* **2009**, *14*, 298–320. [CrossRef] [PubMed]

© 2020 by the authors. Licensee MDPI, Basel, Switzerland. This article is an open access article distributed under the terms and conditions of the Creative Commons Attribution (CC BY) license (http://creativecommons.org/licenses/by/4.0/).

Article

Transformation of Contaminants of Emerging Concern (CECs) during UV-Catalyzed Processes Assisted by Chlorine

Edyta Kudlek

Faculty of Energy and Environmental Engineering, Silesian University of Technology, Konarskiego 18, 44-100 Gliwice, Poland; edyta.kudlek@polsl.pl; Tel.: +48-32-237-24-78

Received: 24 November 2020; Accepted: 6 December 2020; Published: 8 December 2020

Abstract: Every compound that potentially can be harmful to the environment is called a Contaminant of Emerging Concern (CEC). Compounds classified as CECs may undergo different transformations, especially in the water environment. The intermediates formed in this way are considered to be toxic against living organisms even in trace concentrations. We attempted to identify the intermediates formed during single chlorination and UV-catalyzed processes supported by the action of chlorine and hydrogen peroxide or ozone of selected contaminants of emerging concern. The analysis of post-processing water samples containing benzocaine indicated the formation of seven compound intermediates, while ibuprofen, acridine and β-estradiol samples contained 5, 5, and 3 compound decomposition by-products, respectively. The number and also the concentration of the intermediates decreased with the time of UV irradiation. The toxicity assessment indicated that the UV-catalyzed processes lead to decreased toxicity nature of post-processed water solutions.

Keywords: contaminants of emerging concern; UV; advanced oxidation processes; by-products

1. Introduction

Contaminants of emerging concern (CECs) have been discovered in every type of natural water worldwide [1–3]. Christen et al. [4] and Kim et al. [5] reported that the exposure to CECs from the group of pharmaceuticals and pesticides causes a deterioration of human health, including the appearance of cancer diseases and cognitive effects. This proves that these compounds are harmful to aquatic organisms and the entire ecosystem fed by waters containing CECs [6].

The literature indicates that many CECs are hardly degradable or completely resistant to conventional water and wastewater treatment methods such as coagulation, sedimentation, filtration, and biological-based processes [7,8]. Even adsorption on activated carbon cannot cope with the removal of compounds with low hydrophobicity from different water matrices [9]. Compounds with phenolic and amine groups or other electron-donating functional groups can be decomposed by selective oxidants such as ozone (O_3), hydrogen peroxide (H_2O_2), or chlorine and chlorine dioxide [10,11].

Water treatment plants still use the chlorination process as one of the most effective methods for water disinfection. Chlorine (Cl_2), chlorine dioxide (ClO_2), or sodium hypochlorite (NaOCl) reagents are used as disinfectants. However, the oxidizing effect can also be used as an effective and simple method for decomposing organic compounds.

The main oxidizing agent is hypochlorous acid (HOCl), which is formed by the aqueous transformation of Cl_2 according to Equation (1) [12]. HOCl in water solution can dissociate to hypochlorite anions (ClO^-) (Equation (2)). In neutral water conditions, ClO^- are less effective oxidants than HOCl [13].

$$Cl_2 + H_2O \rightarrow HOCl + Cl^- + H^+ \tag{1}$$

$$HOCl \leftrightarrow ClO^- + H^+ \quad (2)$$

The addition of another oxidant or precursor of reactive radicals to the reaction matrix, such as H_2O_2 or O_3, can improve the decomposition of organic compounds in two ways: (1) by the direct action with the compound's molecule or (2) by the initiation of the generation process of other high reactive species [14]. Moreover, implementing the chlorination process together with UV light, can improve the decomposition of contaminant. During UV-catalyzed processes, decomposition of compounds is also caused by their direct photodecomposition and the generation of several radicals (Equations (3) and (4)), like $HO^\bullet$, $O^{-\bullet}$ and reactive chlorine species such as chlorine atoms ($Cl^\bullet$) and ($Cl_2^{\bullet-}$) [15–17]. The oxidation potential of $Cl^\bullet$ and $Cl_2^{\bullet-}$ is 2.47 V and 2.00 V, respectively [18]. Compared to the oxidation potential of $HO^\bullet$, which is equal to 2.80 V, they can also be called strong oxidants. Fang et al. [15] pointed out that $Cl^\bullet$ can react more effectively with acetic acid, benzoic acid, and phenol than $HO^\bullet$.

$$HOCl/OCl^- + h\nu \rightarrow HO^\bullet/O^{\bullet-} + Cl^\bullet \quad (3)$$

$$Cl^\bullet + Cl^- \leftrightarrow Cl_2^{\bullet-} \quad (4)$$

The coexistence of different types of oxidants not always has a positive effect on the decomposition of contaminants. HOCl and ClO^-, which did not come into reaction with contaminates or decompose to reactive radicals, can act as a scavenger for $HO^\bullet$ and $Cl^\bullet$ according to Equations (5)–(8) [15].

$$HO^\bullet + HOCl \rightarrow ClO^\bullet + H_2O \quad (5)$$

$$Cl^\bullet + HOCl \rightarrow + H^+ \quad (6)$$

$$HO^\bullet + OCl^- \rightarrow ClO^\bullet + OH^- \quad (7)$$

$$Cl^\bullet + OCl^- \rightarrow ClO^\bullet + Cl^- \quad (8)$$

The exact transformation mechanisms of different CECs during reactive chlorine species coexisting with other oxidants and UV radiation are still unclear. Oxidants and radicals' action removes CECs or deactivates pathogenic microorganisms and reacts with other compounds present in the disinfected water matrix, leading to several harmful chlorination by-products. The most known chlorination intermediates are trihalomethanes (THMs), and haloacetic acids (HAAs) [19,20], but also chlorination by-products of several CECs are detected in chlorinated water.

The paper presents an attempt to identify the intermediates of selected CECs formed during UV-catalyzed oxidation processes conducted in the presence of chlorine supported by the action of hydrogen peroxide and ozone. Identification of transformation products arising in real water matrices and the determination of their toxicity will give an accurate picture of their potentially dangerous impact on aquatic ecosystems. The comprehension of the CECs decomposition pathways caused by specific physicochemical factors will allow for the development of effective hybrid methods for eliminating the aquatic environment. A single chlorination process was carried out to estimate the decomposition ability of chlorine. The action of chlorine was supported by UV irradiation compared with O_3 and H_2O_2. CECs from the group of pharmaceutical compounds, dye additives, and synthetic hormones were introduced into water solutions based on deionized and surface water and subjected to oxidation agents. The generated by-products were extracted from the post-processed water solutions by the use of Solid Phase Extraction and then chromatographically analyzed and identified based on their mass spectra compared to the National Institute of Standards and Technology NIST v17 database. Bacterial-based toxicological tests confirmed the potentially toxic nature of the intermediates.

2. Results and Discussion

Experiments based on the single chlorination and UV-catalyzed chlorination processes assisted by the action of O_3 and H_2O_2 of CECs water solutions were performed. These processes were assessed both in terms of removing individual compounds and the formation of decomposition by-products and their toxicity to living water organisms.

2.1. Chlorination of CECs

The first stage of the study was devoted to assessing the effectiveness of the CECs' decomposition during dark chamber chlorination, where chlorine was introduced into the water in the form of NaOCl. The influence of chlorine dose on the decomposition of tested CECs in deionized water solutions and surface water solutions is presented in Figures 1 and 2, respectively. The decomposition of compounds occurs during action with HOCl and ClO$^-$ according to Equation (9) [13], leading to the formation of new products, which also should be decomposed.

$$HOCl/OCl^- + \text{organic contaminant} \rightarrow \text{product} \qquad (9)$$

Sivey and Roberts [21] demonstrated that Cl_2 under low pH values and Cl_2O could act as active chlorination agents during the chlorination process and interact with compounds in the treated water solutions. All tested CECs have an aromatic ring in their structure. Therefore, their decomposition should occur due to specific reactions on certain moieties bound to the aromatic ring and the electrophilic substitution of chlorine in the *ortho* or *para* position [22].

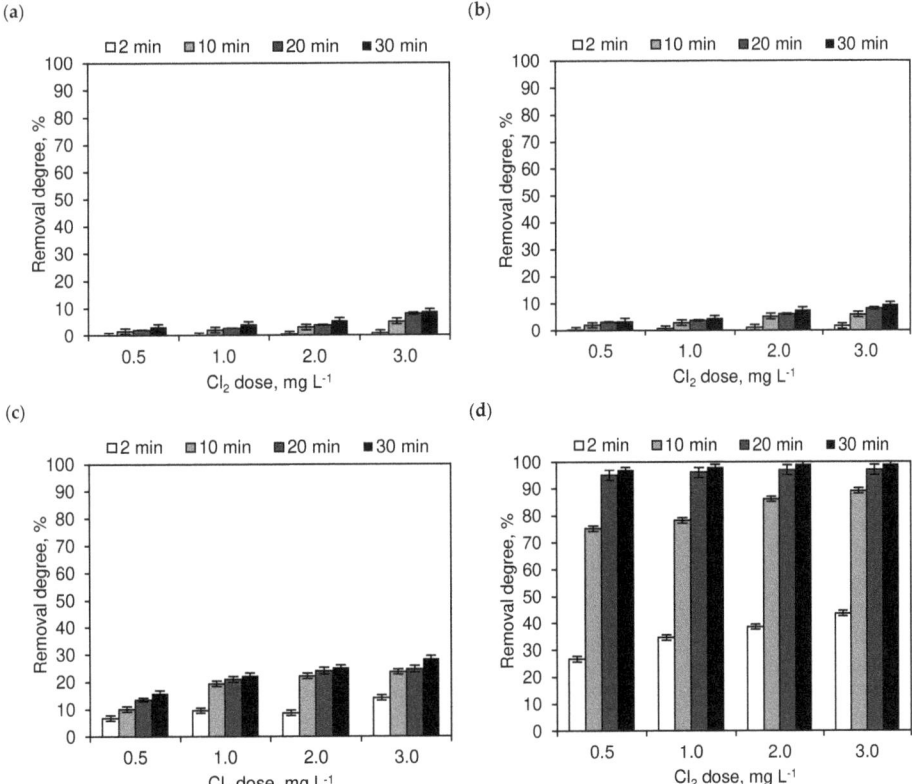

Figure 1. Change of (**a**) IBU, (**b**) BE, (**c**) ACR, and (**d**) E2 concentration during single chlorination of deionized water solutions.

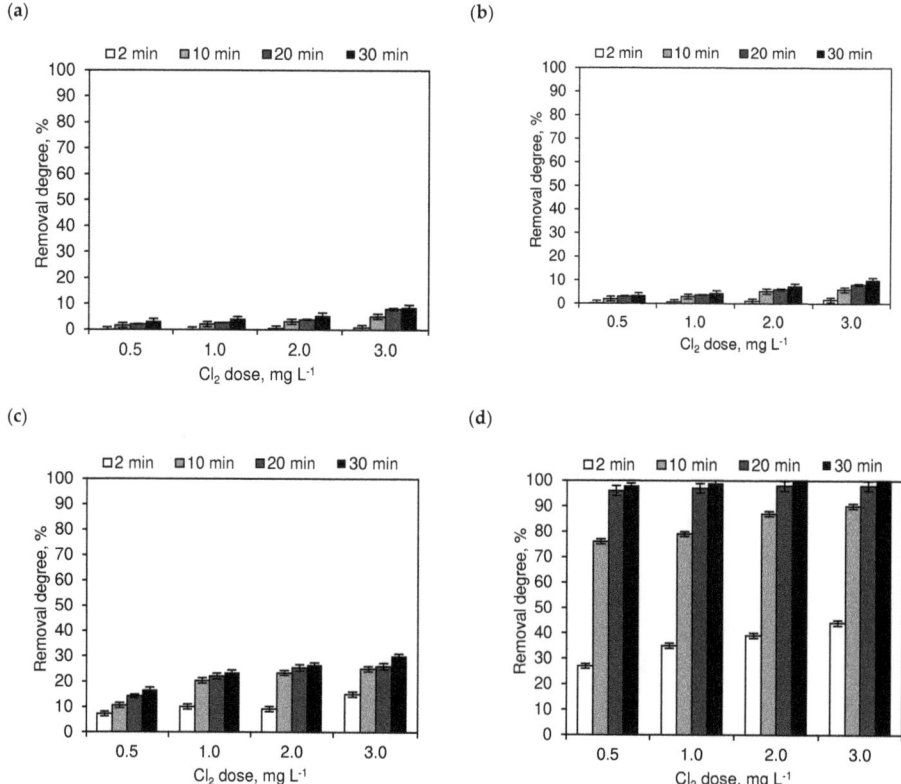

Figure 2. Change of (**a**) IBU, (**b**) BE, (**c**) ACR, and (**d**) E2 concentration during single chlorination of surface water solutions.

As expected, the removal degree of all tested compounds increased with the increase of the chlorine concentration. IBU and BE were the least susceptible to the action of chlorine. No increase of IBU concentration was observed after 2 min reaction time for chlorine doses equal to 0.5 and 1.0 mg L^{-1} in deionized and surface water solutions. The BE removal noted after a 2 min reaction time for the chlorine dose equal to 0.5, 1.0, 2.0, and 3.0 mg L^{-1} was only 0.2, 0.7, 1.1, and 1.6%, respectively, in deionized water solutions. An extension of the reaction time to 20 min by the chlorine dose equal to 0.5 mg L^{-1} allowed for a 3.1% removal of these CECs and an 8% removal for the chlorine content equal to 3.0 mg L^{-1}. Meanwhile, the IBU concentration decreased only by 9% after 30 min of process elongation by the 3.0 mg L^{-1} chlorine dose. Similar low IBU decomposition ability under chlorine action was noted by Xiang et al. [23]. The removal degree of this compound did not exceed 3.1% after 20 min of dark chlorination. The obtained results reconfirmed the recalcitrance of IBP to chlorination.

Higher decomposition rates were noted for ACR. This compound's concentration in the presence of 3.0 mg L^{-1} of chlorine decreased by over 20% after 10 min of single dark chlorination and over 26% after 30 min of process duration. It can be concluded that single NaOCl as a source of HOCl and ClO^- is not sufficient for the decomposition of IBU, BE, and ACR.

A reverse observation was noted in the case of the hormone E2 chlorination. The addition of only 0.5 mg L^{-1} of chlorine to the compound solution based on deionized water led to over 75% and 98% decomposition after 10 and 30 min, respectively. Meanwhile, a complete removal of E2 was observed after 30 min of reaction with the chlorine dose equal to 2 mg L^{-1}. Li et al. [24] also observed a complete removal of this hormone during the chlorination process carried out in neutral conditions.

Similar test results to those obtained for deionized water were noted for a test carried out on surface water (Figure 2). Only the removal degrees of ACR increased for all tested chlorine doses at about 5% and reached, for example, for the free chlorine dose equal to 3.0 mg L^{-1} 30%. Therefore, it can be assumed that organic and inorganic matter in surface water promotes the decomposition of this compound under the influence of chlorine. However, further research in this area is required to determine which component of the natural water matrix is responsible for increasing the removal degree of ACR.

Special attention should be paid to inorganic compounds occurring in real water solutions, i.e., NO_3^- and NH_4^+, whose presence was confirmed in the tested water matrixes. Those compounds can influence the decomposition of compounds and can react with reactive forms of chlorine. However, Qiang and Adams [25] indicated a negligible chlorine reactivity with NH_4^+, although NO_3 and N_2 can be formed during reaction of HOCl with NH_3. This leads to the formation of mono-, di- and trichloramine.

2.2. Decomposition of CECs in UV-Catalyzed Chlorination Processes

The unsatisfactory low removal degree of IBU, BE, and ACR noted during single chlorination forces another treatment process. Therefore, processes integrated the action of chlorination with UV irradiation, additionally supported by the presence of H_2O_2 or O_3 were implemented. Figures 3 and 4 compared the removal degrees of tested CECs noted in deionized and surface water solution exposed to UV irradiation in the presence of NaOCl and H_2O_2 (UV/NaOCl/H_2O_2 process) or O_3 (UV/NaOCl/O_3), respectively.

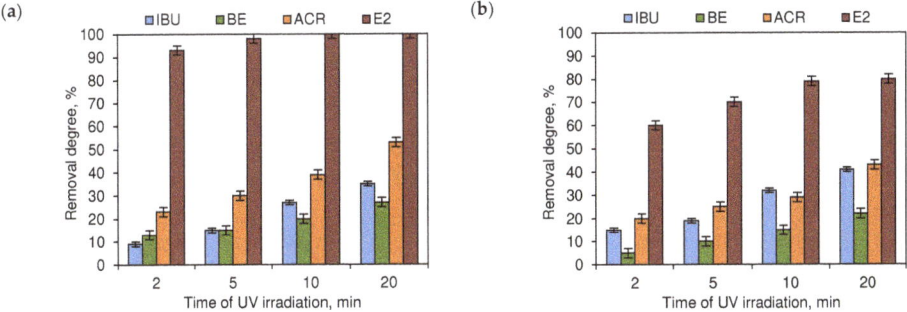

Figure 3. Change of CECs concentration during the UV/NaOCl/H_2O_2 conducted on (**a**) deionized and (**b**) surface water solutions.

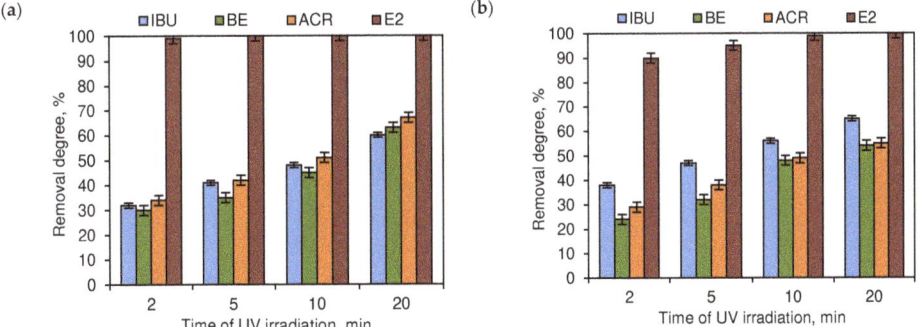

Figure 4. Change of CECs concentration during the UV/NaOCl/O_3 conducted on (**a**) deionized and (**b**) surface water solutions.

The applied UV light source, during the reactions with HOCl, ClO⁻ (Equations (3) and (4)) and O_3 as well as H_2O_2, induced the formation of several reactive radicals which are responsible for a non-selective decomposition of compounds. It should also be emphasized that UV light as an electromagnetic wave carries energy in the form of photons. The interaction of one or more photons with a given compound leads to the chemical transformation of the bonds between atoms that make up the compound molecule [26]. This process results in the photodecomposition of the compound molecule. The maximum absorbance of IBU, BE, and E2 were below the wavelengths, which can get into the irradiated solution. However, those compounds and others occurring, especially in surface water matrixes, can still absorb UV radiation energy and undergo the photo-decomposition process. The energy needed for the dissociation of an H–OH bound, and the formation of HO• exceeds 5 eV [27]. Therefore, the simultaneous application of UV irradiation with different wavelengths and the effect of chlorine and O_3 or H_2O_2 allows for obtaining a required number of free radicals for the decomposition of CECs and their intermediates.

In both UV/NaOCl/H_2O_2 and UV/NaOCl/O_3 processes, preferable removal degrees of BE, ACR, and E2 were observed for compounds occurring in deionized water solutions. This difference in process effectiveness was especially notable during the UV/Cl_2/H_2O_2 process. For example, 2 min of process implementation led to a 60% removal of E2 in the surface water solution, while removing this compound noted in the deionized water matrix exceeded 93%. After 20 min of process duration, complete removal of E2 in deionized water was noted. The concentration of this contaminant in surface water was reduced by 80% (20 min of process duration). The final removal of BE and ACR noted in deionized water samples was equal to 27 and 53%, whereas removing these compounds observed in surface water reached only 22 and 43%, respectively. Only in the case of IBU were higher removal degrees were noted in surface water solutions. For example, after 20 min of UV/NaOCl/H_2O_2 process, the concentration of IBU decreased by 35% in deionized water solutions and over 41% in surface water solutions. Previous studies [28] on the influence of organic and inorganic compounds on the decomposition of IBU in UV-catalyzed processes indicated that the presence of Ca^{2+}, Mg^{2+}, NH_4^+, Cl^-, CO_3^{2-}, HCO_3^-, HPO_4^{2-} as well as SO_4^{2-} ions increased the decomposition process of this pharmaceutical compound.

Higher compound removal degrees were noted during the UV/NaOCl/O_3 process. E2 occurring in deionized water was completely removed after 5 min of UV irradiation, and after 20 min of process implementation, complete removal of this compound was also noted in the surface water matrix. BE and ACR concentration was reduced in deionized water by over 63% and in the surface water only by 55%. Meanwhile, the removal degree of IBU in surface water reached a value of 65%.

It can be concluded that the higher effectiveness of the UV/Cl_2/O_3 against the UV/Cl_2/H_2O_2 process was the result of the formation of a larger number of radicals during the O_3 self-decomposition in water. Among these radicals, HO•, HO_2•, HO_3•, HO_4•, O_2•⁻ and O_3•⁻ can be mentioned [29]. Whereas the irradiation of H_2O_2 with UV light leads to the formation of OH• radicals [30]. However, HO• radicals are endowed with the strongest oxidation potential, and they can abate compounds that are resistant to O_3 or H_2O_2 decomposition [31]. It should also be mentioned that UV irradiation leads to an increase of the quantum yields and the molar absorption coefficients of OCl, OCl, leading to a higher and faster production of radicals during the decomposition processes [15,32,33].

2.3. Identification of Decomposition By-Products

The implementation of the tested CECs decomposition processes allows for a decrease of the contaminant's concentrations and leads to the generation of several by-products. Those intermediates were formed during reactions between the parent compounds and chlorine and/or other reactive species. The intermediates were identified based on their mass spectra using the NIST v17 database (Table 1). The identified by-products decomposed more slowly than the parent micropollutants and were detected even after 20 min of UV-catalyzed process implementation. The single chlorination process led to four BE intermediates: Ethyl 4-chlorobenzoate, 4-Chloroaniline, Chlorohydroquinone, and 2,5-Dichlorohydroquinone, while the chlorination of ACR solutions resulted in the formation of 9-Chloroacridine and Salicylic acid. The subjection of IBU water solutions to single dark chlorination resulted in three intermediates: 1-(4-Isobutylphenyl)ethanol, 4-Acetylbenzoic acid, and 4-Ethylbenzaldehyde. Intermediates detected in each process are summarized in Table 2. The chlorination process led to small modifications in the structures of the compounds and the formation of more oxidized or chlorinated molecules. This is connected because HOCl can react with organic compounds in three types of reactions: oxidation reactions, addition reactions to unsaturated bonds, and electrophilic substitution reactions at nucleophilic sites [34]. Electrophilic substitution reactions are considered the most common mechanisms during chlorination; therefore, chlorine substitution sites will most likely be on the tested compounds' aromatic ring [35].

Table 1. Identified CECs by-products during the performed experiments.

Parent Compound	Identified Compound	Structural Formula	CAS-RN	Similarity, %	Molecular Weight
IBU	1-Hydroxyibuprofen	$C_{13}H_{18}O_3$	53949-53-4	96	222.28
	1-(4-Isobutylphenyl)ethanol	$C_{12}H_{18}O$	40150-92-3	84	178.27
	4'-Isobutylacetophenone	$C_{12}H_{16}O$	38861-78-8	98	176.25
	4-Acetylbenzoic acid	$C_9H_8O_3$	586-89-0	85	164.16
	4-Ethylbenzaldehyde	$C_9H_{10}O$	4748-78-1	92	134.17
BE	Ethyl 4-hydroxybenzoate	$C_9H_{10}O_3$	120-47-8	74	166.17
	Ethyl 4-chlorobenzoate	$C_9H_9ClO_2$	7335-27-5	84	184.62
	4-Chloroaniline	C_6H_6ClN	106-47-8	86	127.57
	4-Chlorophenol	C_6H_5ClO	106-48-9	99	128.55
	3,4-Dichlorophenol	$C_6H_4Cl_2O$	95-77-2	98	163.00
	Chlorohydroquinone	$C_6H_5ClO_2$	615-67-8	75	144.55
	2,5-Dichlorohydroquinone	$C_6H_4Cl_2O_2$	824-69-1	80	179.00
ACR	Acridone	$C_{13}H_9NO$	578-95-0	70	195.22
	Acridine-10-oxide	$C_{13}H_9NO$	10399-73-2	75	195.22
	2-Hydroxyacridine	$C_{13}H_9NO$	22817-17-0	90	195.22
	9-Chloroacridine	$C_{13}H_8ClN$	1207-69-8	72	213.66
	Salicylic acid	$C_7H_6O_3$	69-72-7	80	138.12
E2	2-Hydroxyestradiol	$C_{18}H_{24}O_3$	362-05-0	92	288.40
	Estradiol-3,4-quinone	$C_{18}H_{22}O_3$	144082-88-2	78	286.40
	4-(1-Hydroxyethyl)phenol	$C_8H_{10}O_2$	2380-91-8	80	138.16

Table 2. CECs by-products identified in selected processes during the performed experiments.

Compound	Identified Compound	Cl_2			$UV/Cl_2/H_2O_2$			$UV/Cl_2/O_3$		
		2 min	10 min	20 min	2 min	10 min	20 min	2 min	10 min	20 min
IBU	1-Hydroxyibuprofen	-*	-	-	-	-	-	+	-	-
	1-(4-Isobutylphenyl)ethanol	-	+	+	+	+	-	+	-	-
	4'-Isobutylacetophenone	-	-	-	-	+	+	+	+	+
	4-Acetylbenzoic acid	+	+	+	+	+	+	+	+	+
	4-Ethylbenzaldehyde	+	+	+	-	+	+	+	+	+
BE	Ethyl 4-hydroxybenzoate	-	-	-	+	+	+	+	+	-
	Ethyl 4-chlorobenzoate	-	+	+	+	+	-	+	+	-
	4-Chloroaniline	+	+	+	+	+	-	+	-	-
	4-Chlorophenol	-	-	-	+	+	-	+	+	-
	3,4-Dichlorophenol	-	-	-	+	+	-	+	+	-
	Chlorohydroquinone	+	+	+	+	+	-	+	+	-
	2,5-Dichlorohydroquinone	+	+	+	+	-	-	+	-	-
ACR	Acridone	-	-	-	+	+	+	-	+	+
	Acridine-10-oxide	-	-	-	+	+	+	-	+	+
	2-Hydroxyacridine	-	-	-	+	+	+	-	+	+
	9-Chloroacridine	-	+	+	+	+	-	+	+	-
	Salicylic acid	-	+	+	+	-	-	+	-	-
E2	2-Hydroxyestradiol	-	-	-	+	+	-	+	-	-
	Estradiol-3,4-quinone	-	-	-	+	+	-	+	-	-
	4-(1-Hydroxyethyl)phenol	+	+	+	+	+	+	+	+	-

* - — not detected; + — detected.

During the implemented decomposition processes, the IBU intermediates were mainly formed at the first stage by hydroxylation and chlorine substitution. However, the implemented GC-MS analysis does not allow for detecting IBU by-products with chlorine atoms in their structure. Such compounds were identified by Li et al. [36] and Xiang et al. [23]. The possible decomposition pathway of IBU with the identified intermediates is shown in Figure 5.

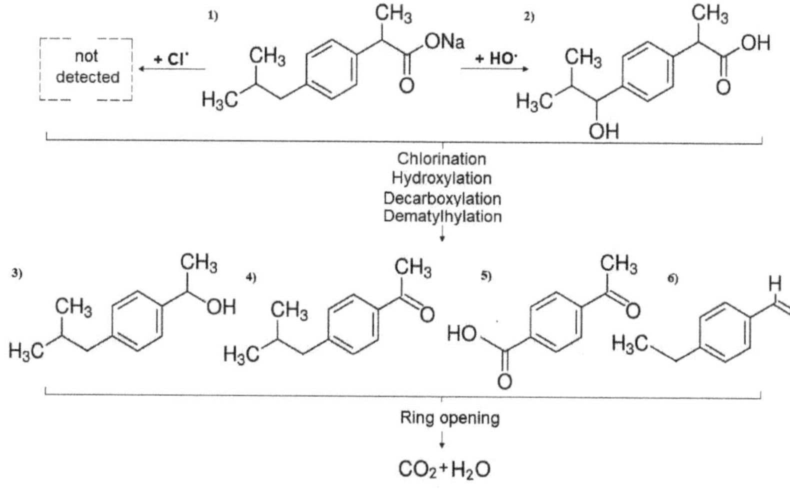

Figure 5. Possible decomposition of (**1**) IBU with the identified intermediates (**2**) 1-Hydroxyibuprofen, (**3**) 1-(4-Isobutylphenyl)ethanol, (**4**) 4′-Isobutylacetophenone, (**5**) 4-Acetylbenzoic acid, and (**6**) 4-Ethylbenzaldehyde.

Ethyl 4-hydroxybenzoate and Ethyl 4-chlorobenzoate were formed by the denitration of the BE molecule and the substitution of the nitric group by the hydroxyl group and chlorine, respectively. Other BE intermediates were possibly generated by the attack of OH• and chlorine on the compound's phenolic ring. The possible decomposition pathway of BE is summarized in Figure 6.

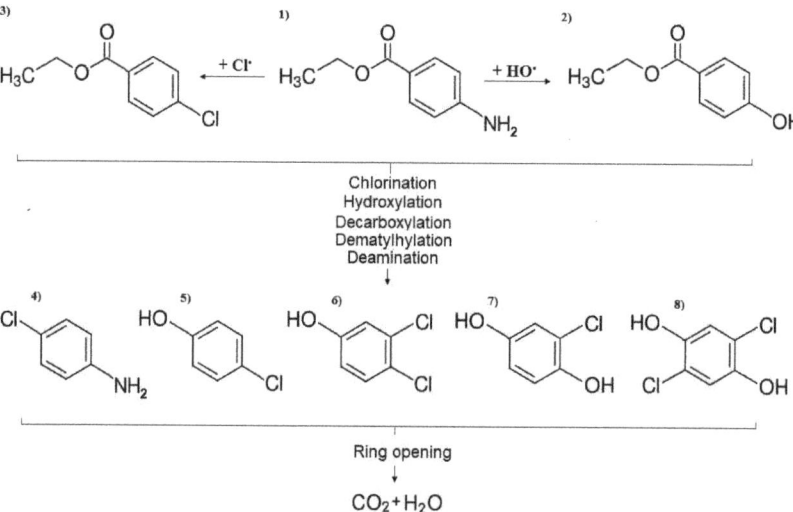

Figure 6. Possible decomposition of (**1**) BE with the identified intermediates (**2**) Ethyl 4-hydroxybenzoate, (**3**) Ethyl 4-chlorobenzoate, (**4**) 4-Chloroaniline, (**5**) 4-Chlorophenol, (**6**) 3,4-Dichlorophenol, (**7**) Chlorohydroquinone, and (**8**) 2,5-Dichlorohydroquinone.

The ACR decomposition by-products Acridone, Acridine-10-oxide, and 2-Hydroxyacridine were mainly formed by the attack of reactive oxygen species of the compound molecule. Meanwhile, 9-Chloroacridine results from the substitution of chlorine to the phenolic ring (Figure 7). Further hydroxylation and the deamination of the formed intermediates led to carbon atom ring-opening and Salicylic acid formation, which was subject to further decomposition.

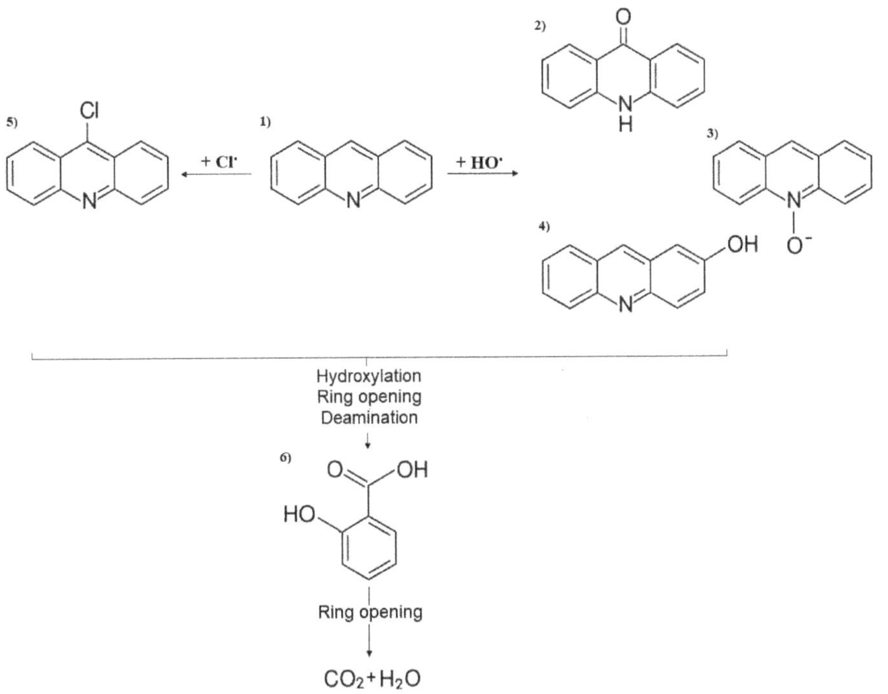

Figure 7. Possible decomposition of (**1**) ACR with the identified intermediates, (**2**) Acridone, (**3**) Acridine-10-oxide, (**4**) 2-Hydroxyacridine, (**5**) 9-Chloroacridine, and (**6**) Salicylic acid.

The applied analytical method based on gas chromatography allows only for the detection of three E2 intermediates, resulting from reactive oxygen species action (Figure 8). Steroid hormones are typically composed of four carbon atoms rings—three cyclohexane rings and one cyclopentane ring. The reactive species, in general, firstly attacks the first cyclohexane ring, which leads to the formation of hydroxylated compounds like 2-Hydroxyestradiol. Li et al. [24] pointed out that the decomposition of E2 occurs by the halogenation of the aromatic ring followed by the cleavage of the benzene moiety and chlorine substitution formation generation of THMs and HAAs from phenolic intermediates.

2.4. Toxicological Assessment

A toxicological assessment is necessary to determine whether the proposed decomposition processes of pollutants do not deteriorate the quality of treated water solutions. It has already been proved that the initial contaminants occurring in non-treated water were sometimes less toxic than the intermediates detected in post-processed water solutions [37]. The preformation of fast toxicological tests like the bacterial-based Micrtox® gives a quick response about the impact of the post-processed water on living organisms. Studies indicated a good interspecies correlation between saltwater bacteria like the used *Aliivibrio fischeri* and other freshwater bacteria or fishes [38]. The *Aliivibrio fischeri* are also considered extremely sensitive to a wide range of pollutants and reagents occurring in water [39].

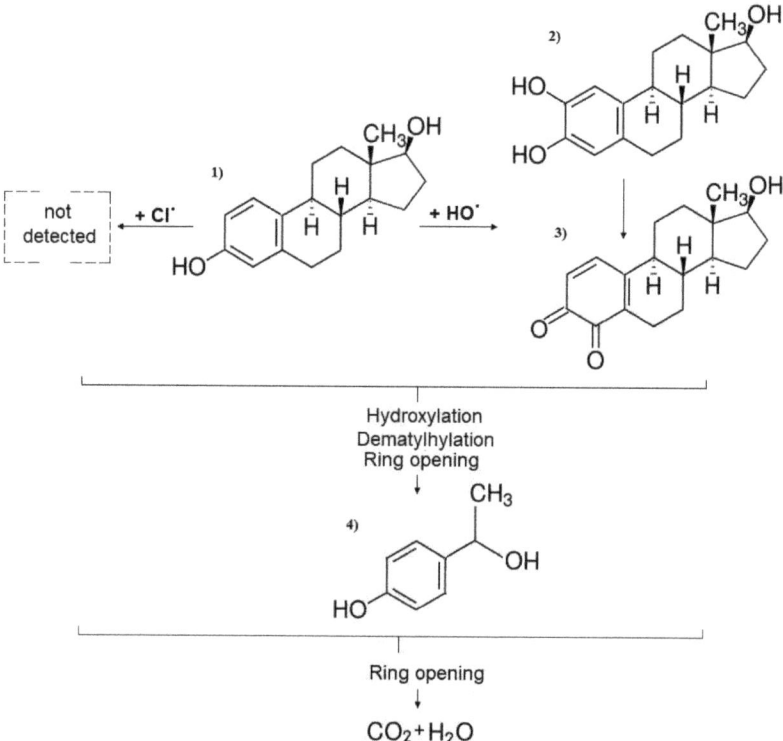

Figure 8. Possible decomposition of (**1**) IBU with the identified intermediates, (**2**) 2-Hydroxyestradiol, (**3**) Estradiol-3,4-quinone, and (**4**) 4-(1-Hydroxyethyl)phenol.

It should be noted that chlorine is used in water treatment processes to protect water before secondary contamination [40]. Therefore, it should be toxic to pathogens and small test organisms. Before the proper toxicological test, the adopted sample preparation methodology allowed for the exclusion of chlorine influence on test bacteria. Chlorinated water samples without the addition of tested CECs were characterized by a toxicological effect lower than 9% (Figure 9), which classified them, according to the guidelines given by Mahugo Santana et al. [41], as non-toxic. Therefore, the results presented in Figures 10–12 resulted from generated parent compound intermediates.

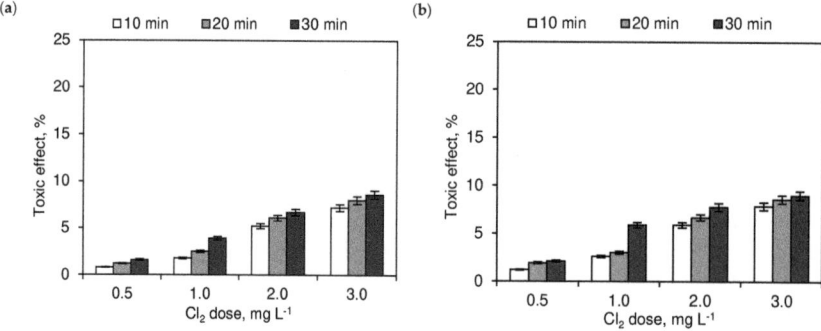

Figure 9. Change in the toxicity of chlorinated (**a**) deionized and (**b**) surface water samples without the addition of CECs.

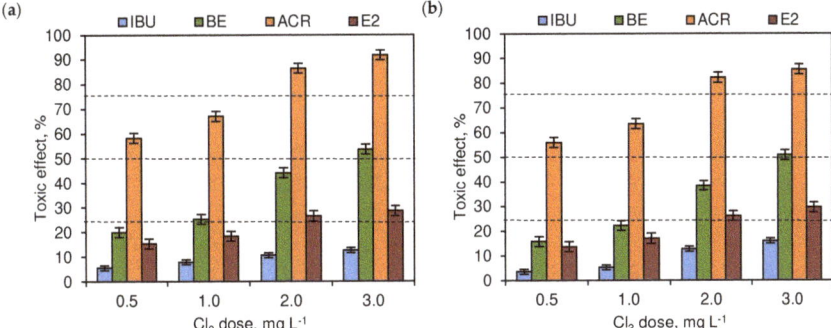

Figure 10. Change in the toxicity of (**a**) deionized water and (**b**) surface water solution after single chlorination (30 min) and the addition of CECs (dashed lines indicate the boundaries between the toxicity classes).

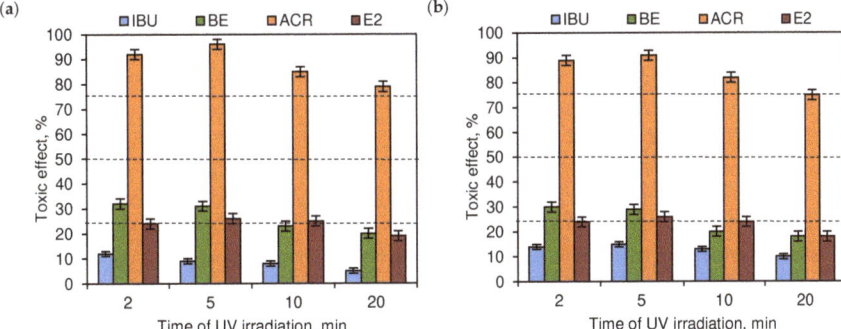

Figure 11. Change in the toxicity of the tested (**a**) deionized water and (**b**) surface water solution after the UV/NaOCl/H_2O_2 process (dashed lines indicate the boundaries between the toxicity classes).

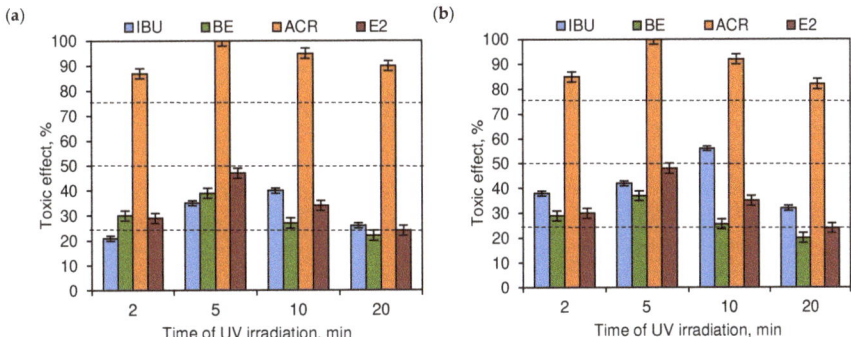

Figure 12. Change in the toxicity of the tested (**a**) deionized water and (**b**) surface water solution after the UV/NaOCl/O_3 process (dashed lines indicate the boundaries between the toxicity classes).

ACR-containing surface water samples before the addition of NaOCl had a toxic effect which exceeded 28%, and according to the toxicity classification, it should be assigned as low toxic (25.0% < toxic effect ≤ 50.0%) against the test bacteria. Whereas IBU, BE, and E2 in the tested concentration of 500 µg L^{-1} were non-toxic (toxic effect ≤ 25.0%).

The presence of chlorine during the implementation of a single chlorination process initiated the compounds' decomposition, leading to an increase in toxicity. It was noted that the toxic effect increased with the increase of the chlorine dose in both deionized and surface water solutions (Figure 10). For example, BE and E2 solutions treated by 0.5 and 1.0 mg L^{-1} of chlorine were still non-toxic, while the dose of 3.0 mg L^{-1} resulted in the increase of the BE solution toxicity to a toxic level (50.0% < toxic effect ≤ 75.0%) and the E2 solution toxicity to a low toxic level. The chlorination process's implementation did not significantly affect the toxicity of the IBU-containing solution, which remained non-toxic for each tested dose of chlorine. The highest toxicity during single chlorination was observed for ACR post-processed solution. The addition of 0.5 or 1.0 mg/L of chlorine to the ACR water solution led to the formation of several toxic intermediates in deionized and surface water solutions, which increased the toxicity to a toxic level. The addition of 2.0 and 3.0 mg/L of chlorine resulted in the generation of highly toxic solutions (toxic effect > 75.0%). The toxicity noted for deionized water solutions containing BE, ACR, and E2 (Figure 10a) was 5%, 8%, and 2%, respectively, higher than the toxicity noted for surface water solutions (Figure 10b).

The post-processed samples obtained during the UV-catalyzed treatment methods were also subjected to toxicological tests (Figures 11 and 12). It was noted that the implementation of UV irradiation supported by the action of 1.0 mg L^{-1} chlorine and H_2O_2 or O_3 resulted in an increase of the toxicity of all tested compound solutions in the first 5 min of the treatment process compared to the single chlorination process. This phenomenon confirms the generation of toxic by-products identified during the chromatographic analysis. In general, the toxicity of samples collected after the process carried out with the presence of O_3 (UV/NaOCl/O_3) was higher than the toxicity of samples subjected to the simultaneous action of UV light, chlorine, and H_2O_2 (UV/NaOCl/H_2O_2).

For example, E2 solutions irradiated for 2, 10, and 20 min in the presence of H_2O_2 were characterized by non-toxicity. Only the sample after 5 min of UV/NaOCl/H_2O_2 was low toxic. Samples of the same contaminant subjected to the UV/NaOCl/O_3 process were classified after 5 min as near medium toxic, but after 20 min, their toxicity was reduced to a non-toxic level. The chromatographic analysis indicated that the signals caused by the formed intermediates after 10 min of UV irradiation become weaker than those after 5 min. This means that after this time of the process, the concentration of by-products decreased. This was also reflected in the toxicity results. All tested compounds' toxicity increased in the first 5 min of UV-catalyzed process implementation and then decreased after 10 and 20 min of process duration. For example, the toxicity of the ACR-containing surface water solution treated by UV/NaOCl/H_2O_2 decreases from a high toxic level to a medium toxic level. Donner et al. [42] noted an increase in the toxicity during the irradiation of carbamazepine solutions with UV light in the first 30 min. ACR is a toxic decomposition by-product of carbamazepine. The occurrence of this compound and other carbamazepine intermediates increased toxicity. Further irradiation of the pharmaceutical solution leads to the decomposition of acridine and decreased solution toxicity.

The BE and E2-containing solutions' toxicity decreased during both UV/NaOCl/H_2O_2 and UV/NaOCl/O_3 processes from a low toxic to a non-toxic level. Only IBU solutions subjected to the UV/NaOCl/O_3 were characterized by increasing toxicity during the first 2, 5, and 10 min. Moreover, the surface water IBU solution after a 10 min exposure to the UV/NaOCl/O_3 process was characterized by a toxic level. It can be summarized that the decomposition of compounds does not always have a beneficial impact on the treated water quality. The generated compound oxidation by-products can radically influence water toxicity and force the necessity to implement further and more complex treatment processes.

3. Materials and Methods

3.1. Water Samples

The research subject constituted water solutions prepared on deionized water witch a conductivity of 18 MΩ cm^{-1} and surface water matrices (conductivity of 0.152 mS cm^{-1}, TOC of 1.546 mg L^{-1}, COD of 89 mg O_2 L^{-1}, N-NH_4 of 0.9 mg L^{-1}, N-NO_3 of 2.1 mg L^{-1}) spiked with CECs standards. Ibuprofen sodium salt (IBU), benzocaine (BE), acridine (ACR), and β-estradiol (E2) were chosen as representative compounds from the group of pharmaceutical compounds, dye additives, and synthetic hormones. The concentration of the compounds in the prepared water solutions was set on 500 µg L^{-1}. Each compound standard solution was prepared by dissolving 10 mg of each analyte in 10 mL of methanol. The use of compound standard solutions allows for the complete dissolution of the analytes and obtains precisely defined CECs concentrations. Standards of the tested CECs with a purity of over 99%, 97%, and 98% were purchased from Sigma-Aldrich (Poznań, Poland). The characteristic of the compound is presented in Table 3. The selected compound concentrations, which exceeded the usual environmental concentrations, allowed for an increase in the accuracy of the analytical measurements. Absorption spectra of all tested compounds were measured using the Spectroquant Pharo 300 UV/Vis spectrophotometer by Merck (Darmstadt, Germany) and compared in Figure 13. The maximum absorbance λ_{max} for IBU, BE, ACR, and E2 was estimated for both deionized and surface water matrixes and were set on 274 nm, 286 nm, 340 nm 280 nm, respectively.

Table 3. Characteristics of the tested organic compounds [43].

Compound	Structural Formula	Molecular Formula	Molecular Weight, g mol^{-1}	Solubility in Water, mg L^{-1}
IBU		$C_{13}H_{17}NaO_2$	228.26	100
BE		$C_9H_{11}NO_2$	165.19	1310
ACR		$C_{13}H_9N$	179.22	38.4
E2		$C_{18}H_{24}O_2$	272.38	3.6

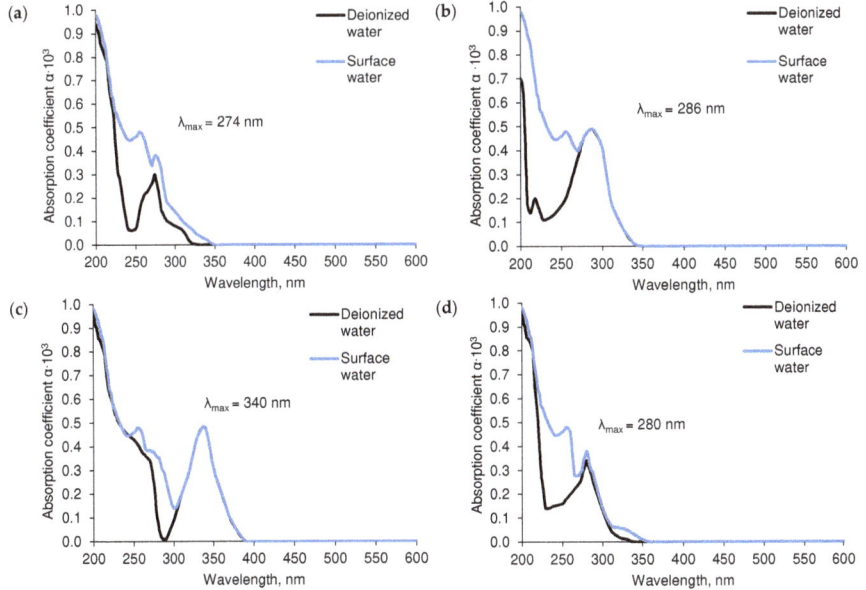

Figure 13. UV-VIS spectrum of (**a**) IBU, (**b**) BE, (**c**) ACR, and (**d**) E2 in deionized and surface water samples.

The experiments for all CECs were performed separately in neutral conditions. The pH of each tested water solution was adjusted to 7.0 using 0.1 mol L^{-1} NaOH or 0.1 mol L^{-1} HCL. Preliminary studies indicated that the very low volumes of the added alkali or acid did not influence the decomposition of the tested micropollutants.

3.2. Decomposition Processes

All prepared water solutions were subjected to the action of the chlorination process. It was carried out using sodium hypochlorite (NaOCl) with a nominal free chlorine content of 6% (*w/v*) purchased from Chemoform (Sosnowiec, Poland). The experiment was conducted on four different chlorine doses equal to 0.5, 1.0, 2.0, and 3.0 mg L^{-1} and measured as a total chlorine concentration by the use of the HI-93414-02 EPA Compliant Turbidity and Free & Total Chlorine Meter by HANNA Instruments Inc. The chlorine doses were selected as part of preliminary studies considering doses used for the chlorination of tap water under normal and special (emergency water pollution) operating conditions. Therefore, these are doses that can be introduced into the water by any water treatment station that uses chlorination as a water disinfection method.

The single chlorination process was carried out in a dark chamber to omit the influence of any light source on the chlorine caused decomposition of tested compounds. The water samples were also exposed to chlorine's action in the presence of UV irradiation supported by hydrogen peroxide (H$_2$O$_2$) or ozone (O$_3$). The H$_2$O$_2$ and O$_3$ dose used in this study was estimated in preliminary tests and set on 3.0 mg L^{-1}. The single chlorination process was carried out 2, 10, 20, and 30 min and stopped by sodium thiosulphate (Na$_2$S$_2$O$_3$) at a dose of 100 mg L^{-1}, which acts as an excess chlorine removing agent. Na$_2$S$_2$O$_3$ with a purity of 98% was purchased from Merck KGaA (Darmstadt, Germany). O$_3$ was generated by an ozonation machine Ozoner FM500 from WRC Multiozon (Gdańsk, Poland) and introduced in the tested water samples using a ceramic diffuser with a height of 25 mm, and a diameter of 12 mm. The O$_3$ concentration in the water solutions was measured using a photometric method on the Spectroquant® by Merc Sp. z o.o. (Warszawa, Poland). The ozonation reaction was stopped after the UV irradiation time by sodium sulfite Na$_2$SO$_3$ at a concentration of 24 mmol L^{-1}. A 150 Watt medium-pressure mercury lap placed in a glass cooling sleeve by Heraeus (Hanau, Germany) was used as the UV light source during all UV-catalyzed decomposition methods. The irradiation time was set as 2, 5, 10, and 20 min, and the radiation flux emitted by the lamp is summarized in Table 4. Figure 14 shows the characteristic of radiation wavelengths emitted by the used UV lamp, and Table 5 shows the energy of light, which reached the water matrix. The energy of light was calculated based on the multiplication of the Planck's constant with the frequency of light.

Experiments for all tested compounds were carried out separately and repeated three times.

Table 4. Physicochemical characteristics of the studied CECs (Data achieved from the supplier of Heraeus UV lamp reactors KENDROLAB Sp. z o.o.).

	Radiation Flux ϕ, W														
Wavelength λ, nm	238/40	254	265	280	297	302	313	334	366	390	405/08	436	492	546	578
Direct lamp radiation	1.0	4.0	1.4	0.7	1.0	1.8	4.3	0.5	6.4	0.1	3.2	4.2	0.1	5.1	4.7
Radiation passing through the sleeve	-	-	-	-	0.1	0.5	2.5	0.4	5.8	0.1	2.9	3.6	0.1	4.6	4.2

Table 5. Energy values of individual wavelengths emitted by the UV lamp.

Wavelength λ, nm	297	302	313	334	366	390	405	406	407	408	436	492	546	578
Energy of light reaching the water matrix, eV	4.17	4.11	3.96	3.71	3.39	3.18	3.06	3.05	3.05	3.04	2.84	2.52	2.27	2.15

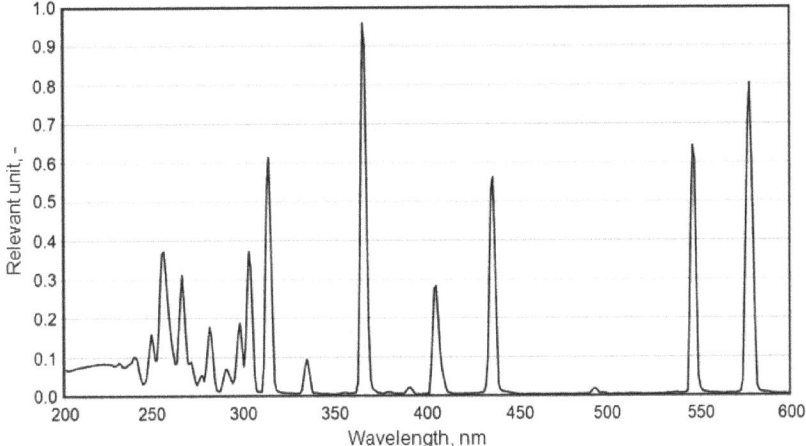

Figure 14. Characteristic of radiation wavelengths emitted by the used UV lamp (Data achieved from the supplier of Heraeus UV lamp reactors KENDROLAB Sp. z o.o.).

3.3. Analytical Procedure and Toxicity Assessment

The analytical procedure of the tested CECs was adopted from previous studies [11] and based on the extraction of analytes from water matrixes by solid-phase extraction (SPE) and their chromatographic analysis. The SPE was performed by the use of Supelclean™ ENVI-8 and Supelclean™ ENVI-18 cartridges obtained from Sigma-Aldrich (Poznań, Poland) with a silica gel base material with C8 (octyl) and silica gel base material with C18 (octadecyl) bonding bed type, respectively. The used SPE procedure allowed for obtaining a recovery of the tested CECs, which exceeded 95%.

The chromatographic analysis was conducted using the GC-MS (EI) chromatograph model 7890B by Perlan Technologies (Warszawa, Poland). The chromatograph was equipped with an SLB™—5 ms 30 m × 0.25 mm capillary column of 0.25 μm film thickness from Sigma-Aldrich (Poznań, Poland). The applied column oven temperature program was: 80 °C (6 min), 5 °C/min up to 260 °C, 20 °C/min up to 300 °C (2 min). Helium 5.0 was used as a carrier gas during the analyses. The temperatures of the ion trap, ion source, and column injector were equal to 150 °C, 230 °C, and 250 °C, respectively. All SPE extract were analyzed twice, in the selected ion monitoring (SIM) mode to monitor CECs concentration and in the total ion current (TIC) mode for the identification of generated compound decomposition by-products. The TIC mode was performed in the range from 50 to 400 m/z.

The percentage of removal of each CECs after the implementation of decomposition processes was calculated according to Equation (10), where C_i and C_p are the initial and post-processed compound concentrations in mg L^{-1}, respectively [44]:

$$\text{Removal (\%)} = \frac{C_i - C_p}{C_i} \cdot 100 \qquad (10)$$

The post-processed samples were also subjected to a spectroscopic analysis performed on the Spectroquant Pharo 300 UV/Vis spectrophotometer by Merck (Darmstadt, Germany), which can measure the UV-VIS spectrum of samples in the range from 200 to 600 nm. The spectroscopic measurement indicated only the decrease of the initial compound concentration (it was discussed in Sections 2.1 and 2.2. based on the percentage of removal calculated from the data obtained by the GC-MS analysis). The UV-VIS spectrum did not indicate the formation of intermediates, which was related to their low concentrations. Therefore, the results were not presented in the paper.

The toxicological evaluation, which gives an answer about the potentially toxic nature of the newly generated decomposition by-products, was carried out using the Microtox® test. The test procedure is based on the measurement of the changes in the behavior of saltwater bioluminescent bacteria *Aliivibrio fischeri* according to the Screening Test procedure of MicrotoxOmni system, which controls the work of the Microtox analyzer Model 500 by Modern Water (London, United Kingdom). The test results were expressed as a percentage of bacterial bioluminescence inhibition caused by changes in the bioindicators' metabolic processes exposed to a toxicant for 15 min. The results were compared to a reference nontoxic sample (2% NaCl solution). The collected samples were subjected to toxicological test after 24 h incubation in a cooled dark chamber to exclude the possible chlorine impact, which was left in the post-processed water solutions.

Assignment errors marked on figures presented in this paper were estimated based on the standard deviation for three repetitions of each test. The error values for all tested samples did not exceed 2.0%.

4. Conclusions

Based on the conducted compound decomposition assessments in the selected oxidation processes, it can be concluded that the lowest compound removal degrees were observed during single dark chlorination. Moreover, the decomposition rates of all tested compounds obtained in this process conducted in deionized water matrixes were very similar to those noted in surface water matrixes. The difference in the compound's removal degrees in both matrices did not exceed 2% and was within the measurement error. Further, the occurrence of chlorine in the reaction matrixes leads to the generation of intermediates with chlorine atoms in their structure. It was also noted that the compound concentration decreases with the increase of the chlorine concentration. Higher chlorine concentrations also lead to an increase of the by-products forming. The conduction of UV-catalyzed oxidation processes shows that the combination of the UV radiation with the action of chlorine and O_3 was more effective for compound decomposition than the UV chlorination process supported by the presence of H_2O_2. The implementation of the UV radiation in oxidants' presence results in the decrease of the number and concentration of formed by-products after 20 min of process elongation. However, it should be noted that during the first 10 min of both UV/NaOCl/H_2O_2 and UV/NaOCl/O_3 processes, several oxidized intermediates were formed. The UV-catalyzed processes also lead to the decrease of the toxicity of the post-processed water solutions, which still depend on the type of decomposed compounds and the UV irradiation time.

Funding: The studies were performed within the project's framework founded by the Polish Ministry of Science and Higher Education grant no. 08/040/BKM20/0138.

Conflicts of Interest: The authors declare no conflict of interest.

References

1. aus der Beek, T.; Weber, F.A.; Bergmann, A.; Hickmann, S.; Ebert, I.; Hein, A.; Küster, A. Pharmaceuticals in the environment-Global occurrences and perspectives. *Environ. Toxicol. Chem.* **2016**, *35*, 823–835. [CrossRef] [PubMed]
2. Grung, M.; Lin, Y.; Zhang, H.; Steen, A.O.; Huang, J.; Zhang, G.; Larssen, T. Pesticide levels and environmental risk in aquatic environments in China—A review. *Environ. Int.* **2015**, *81*, 87–97. [CrossRef] [PubMed]
3. K'oreje, K.O.; Okoth, M.; Van Langenhove, H.; Demeestere, K. Occurrence and treatment of contaminants of emerging concern in the African aquatic environment: Literature review and a look ahead. *J. Environ. Manag.* **2020**, *254*, 109752. [CrossRef]
4. Christen, V.; Hickmann, S.; Rechenberg, B.; Fent, K. Highly active human pharmaceuticals in aquatic systems: A concept for their identification based on their mode of action. *Aquat. Toxicol.* **2010**, *96*, 167–181. [CrossRef] [PubMed]
5. Kim, K.H.; Kabir, E.; Jahan, S.A. Exposure to pesticides and the associated human health effects. *Sci. Total Environ.* **2017**, *575*, 525–535. [CrossRef]
6. Bolong, N.; Ismail, A.F.; Salim, M.R.; Matsuura, T. A review of the effects of emerging contaminants in wastewater and options for their removal. *Desalination* **2009**, *239*, 229–246. [CrossRef]

7. Rizzo, L.; Gernjak, W.; Krzeminski, P.; Malato, S.; McArdell, C.S.; Sanchez Perez, J.A.; Schaar, H.; Fatta-Kassinos, D. Best available technologies and treatment trains to address current challenges in urban wastewater reuse for irrigation of crops in EU countries. *Sci. Total Environ.* **2020**, *710*, 136312. [CrossRef] [PubMed]
8. Falas, P.; Wick, A.; Castronovo, S.; Habermacher, J.; Ternes, T.A.; Joss, A. Tracing the limits of organic micropollutant removal in biological wastewater treatment. *Water Res.* **2016**, *95*, 240–249. [CrossRef]
9. Huerta-Fontela, M.; Galceran, M.T.; Ventura, F. Occurrence and removal of pharmaceuticals and hormones through drinking water treatment. *Water Res.* **2011**, *45*, 1432–1442. [CrossRef] [PubMed]
10. Lee, Y.; von Gunten, U. Oxidative transformation of micropollutants during municipal wastewater treatment: Comparison of kinetic aspects of selective (chlorine, chlorine dioxide, ferrateVI, and ozone) and non-selective oxidants(hydroxyl radical). *Water Res.* **2010**, *44*, 555–566. [CrossRef]
11. Kudlek, E. Decomposition of contaminants of emerging concern in advanced oxidation processes. *Water* **2018**, *10*, 955. [CrossRef]
12. Rott, E.; Kuch, B.; Lange, C.; Richter, P.; Kugele, A.; Minke, R. Removal of Emerging Contaminants and Estrogenic Activity from Wastewater Treatment Plant Effluent with UV/Chlorine and UV/H_2O_2 Advanced Oxidation Treatment at Pilot Scale. *Int. J. Environ. Res. Public Health* **2018**, *15*, 935. [CrossRef] [PubMed]
13. Deborde, M.; von Gunten, U. Reactions of chlorine with inorganic and organic compounds during water treatment—Kinetics and mechanisms: A critical review. *Water Res.* **2008**, *42*, 13–51. [CrossRef] [PubMed]
14. Kudlek, E.; Dudziak, M. Toxicity and degradation pathways of selected micropollutants in water solutions during the O_3 and O_3/H_2O_2 process. *Desalin. Water Treat.* **2018**, *117*, 88–100. [CrossRef]
15. Fang, J.; Fu, Y.; Shang, C. The roles of reactive species in micropollutant degradation in the UV/free chlorine system. *Environ. Sci. Technol.* **2014**, *48*, 1859–1868. [CrossRef]
16. Li, M.; Xu, B.; Liungai, Z.; Hu, H.-Y.; Chen, C.; Qiao, J.; Lu, Y. The removal of estrogenic activity with UV/chlorine technology and identification of novel estrogenic disinfection by-products. *J. Hazard. Mater.* **2016**, *307*, 119–126. [CrossRef]
17. Yang, X.; Sun, J.; Fu, W.; Shang, C.; Li, Y.; Chen, Y.; Gan, W.; Fang, J. PPCP degradation by UV/chlorine treatment and its impact on DBP formation potential in real waters. *Water Res.* **2016**, *98*, 309–318. [CrossRef]
18. Hirakawa, T.; Nosaka, Y. Properties of $O_2^{\bullet-}$ and $OH^{\bullet}$ formed in TiO_2 aqueous suspensions by photocatalytic reaction and the influence of H_2O_2 and some ions. *Langmuir* **2002**, *18*, 3247–3254. [CrossRef]
19. Legay, C.; Leduc, S.; Dubé, J.; Levallois, P.; Rodriguez, M.J. Chlorination by-product levels in hot tap water: Significance and variability. *Sci. Total Environ.* **2019**, *651*, 1735–1741. [CrossRef]
20. Villanueva, C.M.; Cordier, S.; Font-Ribera, L.; Salas, L.A.; Levallois, P. Overview of disinfection by-products and associated health effects. *Curr. Environ. Health Rep.* **2015**, *2*, 107–115. [CrossRef]
21. Sivey, J.D.; Roberts, A.L. Assessing the reactivity of free chlorine constituents Cl_2, Cl_2O, and HOCl toward aromatic ethers. *Environ. Sci. Technol.* **2012**, *46*, 2141–2147. [CrossRef] [PubMed]
22. Núñez-Gaytán, A.M.; Vera-Avila, L.E.; De Llasera, M.G.; Covarrubias-Herrera, R. Speciation and transformation pathways of chlorophenols formed from chlorination of phenol at trace level concentration. *J. Environ. Sci. Health Part A* **2010**, *45*, 1217–1226. [CrossRef] [PubMed]
23. Xiang, Y.; Fang, J.; Shang, C. Kinetics and pathways of ibuprofen degradation by the UV/chlorine advanced oxidation process. *Water Res.* **2016**, *90*, 301–308. [CrossRef] [PubMed]
24. Li, C.; Dong, F.; Crittenden, J.C.; Luo, F.; Chen, X.; Zhao, T. Kinetics and mechanism of 17β-estradiol chlorination in a pilot-scale water distribution systems. *Chemosphere* **2017**, *178*, 73–79. [CrossRef] [PubMed]
25. Qiang, Z.; Adams, C. Determination of monochloramine formation rate constants with stopped-flow spectrometry. *Environ. Sci. Technol.* **2004**, *38*, 1435–1444. [CrossRef]
26. Speight, J.G. *Reaction Mechanisms in Environmental Engineering*; Butterworth-Heinemann: Oxford, UK, 2018; pp. 231–267.
27. Tomanová, K.; Precek, M.; Múčka, V.; Vyšín, L.; Juha, L.; Čuba, V. At the crossroad of photochemistry and radiation chemistry: Formation of hydroxyl radicals in diluted aqueous solutions exposed to ultraviolet radiation. *Phys. Chem. Chem. Phys.* **2017**, *19*, 29402–29408. [CrossRef] [PubMed]
28. Kudlek, E.; Dudziak, M.; Bohdziewicz, J. Influence of Inorganic Ions and Organic Substances on the Degradation of Pharmaceutical Compound in Water Matrix. *Water* **2016**, *8*, 532. [CrossRef]
29. da Silva, L.M.; Wilson, J.F. Trends and strategies of ozone application in environmental problems. *Química Nova* **2006**, *29*, 310–317. [CrossRef]

30. Mierzwa, J.C.; Rodrigues, R.; Teixeira, A.C.S.C. UV-Hydrogen Peroxide Processes. In *Advanced Oxidation Processes for Wastewater Treatment: Emerging Green Chemical Technology*; Academic Press: Cambridge, MA, USA, 2018; pp. 13–48.
31. Bourgin, M.; Borowska, E.; Helbing, J.; Hollender, J.; Kaiser, H.-P.; Kienle, C.; McArdell, C.S.; Simon, E.; von Gunten, U. Effect of operational and water quality parameters on conventional ozonation and the advanced oxidation process O_3/H_2O_2: Kinetics of micropollutant abatement, transformation product and bromate formation in a surface water. *Water Res.* **2017**, *122*, 234–245. [CrossRef]
32. Wang, D.; Bolton, J.R.; Hofmann, R. Medium pressure UV combined with chlorine advanced oxidation for trichloroethylene destruction in a model water. *Water Res.* **2012**, *46*, 4677–4686. [CrossRef]
33. Jin, J.; El-Din, M.G.; Bolton, J.R. Assessment of the UV/Chlorine process as an advanced oxidation process. *Water Res.* **2011**, *45*, 1890–1896. [CrossRef] [PubMed]
34. Cheng, H.; Song, D.; Chang, Y.; Liu, H.; Qu, J. Chlorination of tramadol: Reaction kinetics, mechanism and genotoxicity evaluation. *Chemosphere* **2015**, *141*, 282–289. [CrossRef] [PubMed]
35. Soufan, M.; Deborde, M.; Legube, B. Aqueous chlorination of diclofenac: Kinetic study and transformation products identification. *Water Res.* **2012**, *46*, 3377–3386. [CrossRef] [PubMed]
36. Li, X.; Wang, Y.; Yuan, S.; Li, Z.; Wang, B.; Huang, J.; Deng, S.; Yu, G. Degradation of the anti-inflammatory drug ibuprofen by electro-peroxone process. *Water Res.* **2014**, *63*, 81–93. [CrossRef] [PubMed]
37. Wacławek, S.; Lutze, H.V.; Grübel, K.; Padil, V.V.T.; Černík, M.; Dionysiou, D.D. Chemistry of persulfates in water and wastewater treatment: A review. *Chem. Eng. J.* **2017**, *330*, 44–62. [CrossRef]
38. Zhang, X.J.; Qin, H.W.; Su, L.M.; Qin, W.C.; Zou, M.Y.; Sheng, L.X.; Zhao, Y.H.; Abraham, M.H. Interspecies correlations of toxicity to eight aquatic organisms: Theoretical considerations. *Sci. Total Environ.* **2010**, *408*, 4549–4555. [CrossRef]
39. Marugán, J.; Bru, D.; Pablos, C.; Catalá, M. Comparative evaluation of acute toxicity by Vibrio fischeri and fern spore based bioassays in the follow-up of toxic chemicals degradation by photocatalysis. *J. Hazard. Mater.* **2012**, *213–214*, 117–122.
40. Du, Y.; Lv, X.-T.; Wu, Q.-Y.; Zhang, D.-Y.; Zhou, Y.-T.; Peng, L.; Hu, H.-Y. Formation and control of disinfection byproducts and toxicity during reclaimed water chlorination: A review. *J. Environ. Sci.* **2017**, *58*, 51–63. [CrossRef]
41. Mahugo Santana, C.; Sosa Ferrera, Z.; Torres Padron, M.E.; Santana Rodríguez, J.J. Methodologies for the extraction of phenolic compounds from environmental samples: New Approaches. *Molecules* **2009**, *14*, 298–320. [CrossRef]
42. Donner, E.; Kosjek, T.; Qualmann, S.; Kusk, K.O.; Heath, E.; Revitt, D.M.; Ledin, A.; Andersen, H.R. Ecotoxicity of carbamazepine and its UV photolysis transformation products. *Sci. Total Environ.* **2013**, *443*, 870–876. [CrossRef] [PubMed]
43. Kim, S.; Thiessen, P.A.; Bolton, E.E.; Chen, J.; Fu, G.; Gindulyte, A.; Han, L.; He, J.; He, S.; Shoemaker, B.A.; et al. PubChem Substance and Compound databases. *Nucleic Acids Res.* **2016**, *44*, D1202–D1213. [CrossRef] [PubMed]
44. Homaeigohar, S.; Zillohu, A.U.; Abdelaziz, R.; Hedayati, M.K.; Elbahri, M. A Novel Nanohybrid Nanofibrous Adsorbent for Water Purification from Dye Pollutants. *Materials* **2016**, *9*, 848. [CrossRef] [PubMed]

Publisher's Note: MDPI stays neutral with regard to jurisdictional claims in published maps and institutional affiliations.

© 2020 by the author. Licensee MDPI, Basel, Switzerland. This article is an open access article distributed under the terms and conditions of the Creative Commons Attribution (CC BY) license (http://creativecommons.org/licenses/by/4.0/).

Article

Cleaner Production of Epoxidized Cooking Oil Using A Heterogeneous Catalyst

Maria Kurańska * and Magdalena Niemiec

Department of Chemistry and Technology of Polymers, Cracow University of Technology, Warszawska 24, 31-155 Kraków, Poland; mniemiec@pk.edu.pl
* Correspondence: maria.kuranska@pk.edu.pl; Tel.: +48-126282747

Received: 15 October 2020; Accepted: 28 October 2020; Published: 30 October 2020

Abstract: A cleaner solvent-free process of used cooking oil epoxidation has been developed. The epoxidation reactions were carried out using "in situ"-formed peroxy acid. A variety of ion exchange resins with different cross-linking percentages and particle sizes such as Dowex 50WX2 50-100, Dowex 50WX2 100-200, Dowex 50WX2 200-400, Dowex 50WX4 50-100, Dowex 50WX4 100-200, Dowex 50WX4 200-400, Dowex 50WX8 50-100, Dowex 50WX8 100-200, Dowex 50WX8 200-400 were used in the synthesis as heterogeneous catalysts. No significant effect of the size as well as porosity of the catalysts on the properties of the final products was observed. In order to develop a more economically beneficial process, a much cheaper heterogeneous catalyst—Amberlite IR-120—was used and the properties of the epoxidized oil were compared with the bio-components obtained in the reaction catalyzed by the Dowex resins. The epoxidized waste oils obtained in the experiments were characterized by epoxy values in the range of 0.32–0.35 mol/100 g. To reduce the amount of waste, the reusability of the ion exchange resin in the epoxidation reaction was studied. Ten reactions were carried out using the same catalyst and each synthesis was monitored by determination of epoxy value changes vs. time of the reactions. It was noticed that in the case of the reactions where the catalyst was reused for the third and fourth time the content of oxirane rings was higher by 8 and 6%, respectively, compared to the reaction where the catalyst was used only one time. Such an observation has not been reported so far. The epoxidation process with catalyst recirculation is expected to play an important role in the development of a new approach to the environmentally friendly solvent-free epoxidation process of waste oils.

Keywords: ion exchange resins; waste cooking oil; reuse of catalyst; epoxidation; Circular Economy

1. Introduction

The increasing price of petrochemical raw materials, their limited availability and the growing problem of environmental pollution draw the attention of the chemical industry to sustainable development. One of its assumptions is searching for new renewable raw materials that can be successfully used in the synthesis of chemical compounds. An example of such raw materials is vegetable oils, which in terms of chemical structure consist of triglycerides, i.e., esters of glycerol and three fatty acids, mainly unsaturated [1–3]. Waste vegetable oils are also an interesting raw material [3–6]. Syntheses based on such materials are a more ecological solution owing to the possibility of managing waste generated during the frying process. Such an approach implements the requirements of Circular Economy.

Double bonds in fatty acid chains are reactive sites that allow chemical modifications of vegetable oils to increase their possible applications. One of such modifications is the epoxidation process. During this process unsaturated bonds are oxidized to epoxy groups, which are also called oxirane rings [1]. Among several epoxidation methods, the most important process is the use of carboxylic peracids as an oxidizing agent. Industrially, peracetic acid is the most commonly used material, although the process can also be carried out using performic acid, perfluoroacetic acid, perbenzoic acid, m-chloroperbenzoic acid, and m-nitroperbenzoic acid [7]. Carboxylic peracid is formed by the reaction of a proper organic acid with hydrogen peroxide in the presence of a catalyst, usually in the form of strong mineral acids, acidic ion exchange resins (AIER), or enzymes [8].

The process of epoxidation can be carried out in one or two stages. The one-step method is often called in situ epoxidation. In this process, all components, i.e., vegetable oil, organic acid, hydrogen peroxide and catalyst, are mixed in one reaction vessel. In the two-step method, two reactors are used. In the first one, peracid is obtained and then placed in the second vessel where the main epoxidation of the vegetable oil is carried out [9].

During the epoxidation process of vegetable oils, side reactions may occur, especially when the process is carried out in the presence of strongly acidic catalysts. The type of catalyst has a significant impact on the epoxidation process. On an industrial scale, homogeneous catalysts, mainly strong mineral acids, are most commonly used. The use of such compounds leads to a reduction of the process costs and allows obtaining epoxidized oils with a low content of unsaturated bonds [10]. Examples of mineral acids used in the epoxidation process of vegetable oils, both fresh and waste, are H_2SO_4, HNO_3, H_3PO_4, HCl. It has been found that among these catalysts, H_2SO_4 is the most efficient and effective [7,11,12]. However, the disadvantage of homogeneous catalysts is the relatively low selectivity of converting unsaturated bonds into epoxy groups. This effect is caused by the occurrence of side reactions, especially the oxirane ring-opening reaction, which are catalyzed by strong mineral acids. Moreover, the process in the presence of these catalysts leads to considerable amounts of waste water that is difficult to purify [10].

Improvement of process selectivity, by reducing side reactions, can be achieved using heterogeneous catalysts such as acidic ion exchange resins [13]. The most commonly used resins are Amberlyst 15, Amberlite IR-120, and Dowex 50WX2. These catalysts are copolymers of styrene and divinylbenzene and differ in the form and content of the cross-linking agent. It has been proved that the type of the ion exchange resin that is used has an impact on the epoxidation process of vegetable oils [14]. In addition, increasing the concentration of the heterogeneous catalyst usually results in increasing the conversion of unsaturated bonds into epoxide groups [15–18]. The advantages of using acidic ion exchange resins include also the ease of separating them from the finished product and the possibility of re-use. Even after several uses, the loss of activity of this type of catalyst is observed to be insignificant and the efficiency of the process is slightly reduced [15,19,20].

The oxidation of double bonds in fatty acid chains can also be carried out in the presence of enzymatic catalysts, thanks to which the epoxidation process of vegetable oils is more environmentally friendly. What is more, this method allows reducing the number of side reactions and achieve high efficiency and selectivity of the process [21–23]. However, enzymes are characterized by low stability and their activity decreases with an increasing temperature [23]. In addition, re-use of enzymes for catalytic purposes results in a significant reduction in the efficiency of the epoxidation reaction [21].

Temperature also has an important influence on the process of epoxidation of vegetable oils. According to the literature, as the reaction temperature increases, the time needed to achieve high efficiency decreases [24]. However, too high a temperature intensifies side reactions, mainly the opening of oxirane rings [24].

Other factors that affect the epoxidation reaction include the type of oxidizing agent [7], the molar ratio of reactants [8] and the intensity of mixing [25]. Epoxy compounds derived from vegetable and waste oils have many applications. They can be used directly as plasticizers and stabilizers for plastics [26,27]. Epoxy oils are also a raw material for the preparation of many chemical compounds, such as alcohols, glycols, olefinic and carbonyl compounds, epoxy resins, polyesters, or polyurethanes.

This paper reports on the epoxidation of waste oil from a local restaurant with peroxyacetic acid formed in situ from acetic acid and hydrogen peroxide in the presence of ion exchange resin as a heterogenous catalyst. Emphasis was mainly put on determining the process conditions that are consistent with cleaner production. The following four main aspects were taken into account: easy removal of the catalyst, an inexpensive catalyst ensuring adequate product properties, the lowest possible catalyst concentration ensuring adequate conversion in a relatively short epoxidation reaction time, the possibility of reusing the catalyst. The conversion of waste oil into useful chemicals has attracted significant attention in the fields of green and sustainable chemistry and has prompted the implementation of the Circular Economy rules in the polymer technology.

2. Results and Discussion

The experimentally determined initial iodine number of the used cooking oil was 104 gI_2/100 g meaning 0.41 mol of double bonds per 100 g of the used cooking oil. Acidic Ion Exchange Resin (AIER) is an insoluble gel type catalyst in the form of small yellowish organic polymer beads. AIER offers considerable advantages over conventional chemical methods of epoxidation of vegetable oil by improving the selectivity and reducing undesirable side reactions to a certain level [1,5]. In order to determine the most favorable conditions of used cooking oil epoxidation, the effects of different reaction modifications were studied as described further. Firstly, the effect of the content of a heterogenous catalyst on the epoxidation of waste oil was analyzed. The effect of the catalyst concentration on the in situ epoxidation expressed by the changes of the epoxy value over time is presented in Figure 1.

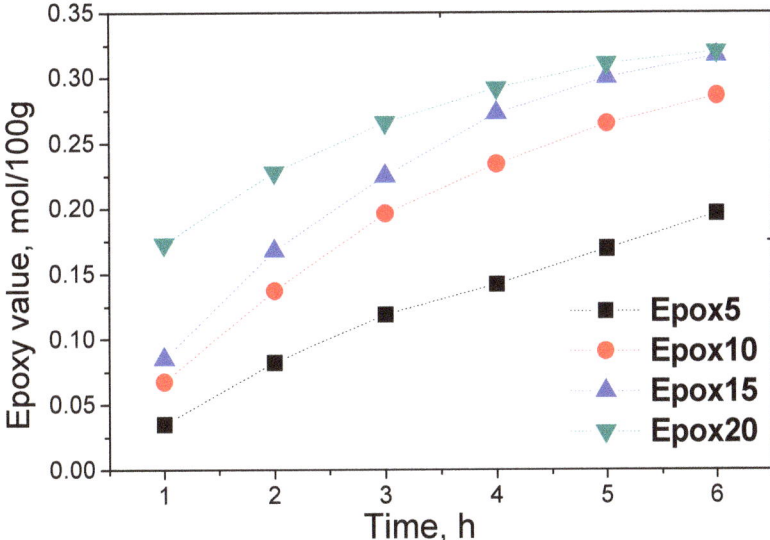

Figure 1. Effect of increasing catalyst concentration on epoxy value of epoxidized used cooking oil.

The oxirane rings content increased monotononically with the reaction time for all reactions. As expected, the efficiency of the reaction is greater when the catalyst concentration is higher. It was observed that after 6 h of the reaction the epoxy values of Epox15 and Epox20 are similar and it is not necessary to increase the catalyst concentration up to 20%. The increase in the intensity of the absorption band during the reaction, characteristic of epoxy groups depending on the catalyst concentration, is presented in Figure 2.

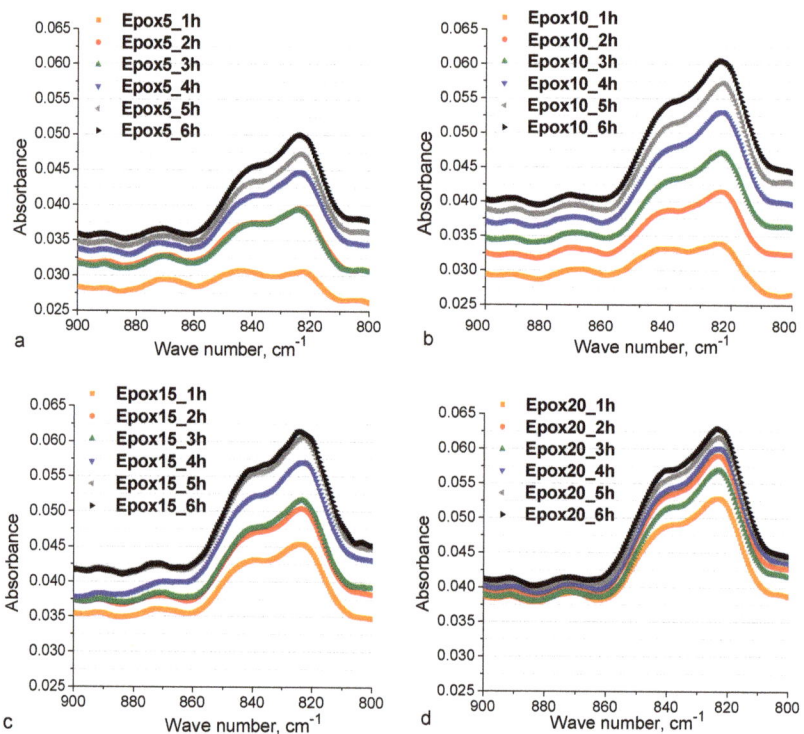

Figure 2. Dependence of absorption band characteristic of epoxy groups on catalyst concentration. (**a**) 5%, (**b**) 10%, (**c**) 15%, and (**d**) 20%.

For Epox5, the peaks in the spectrum exhibit lower intensities compared to the corresponding Epox15 (Figure 2c) and Epox20 (Figure 2d) peaks. The absorbance band intensities correlate with the epoxy number values.

As can be seen in Table 1, the iodine value of the epoxidized used cooking oil confirms a successful epoxidation of the double bonds of the waste oil. Epox20 was characterized by the highest epoxy value and in consequence the lowest iodine value. For all reactions, conversion, efficiency, and selectivity were determined. The epoxy numbers of Epox15 and Epox20 are comparable. Epox15 was characterized by higher selectivity than Epox20, which is associated with a stronger tendency for side reactions to occur when a higher concentration of the catalyst was used.

Table 1. Characteristics of epoxidized UCOs.

Sample	Ev, mol/100 g	Iv, gI$_2$/100 g	Hv, mgKOH/g	C, %	E, %	S, %	Mn, g/mol	Mw, g/mol	D	η, mPa·s
UCO	-	104.0	-	-	-	-	915	921	1.01	68
Epox5	0.241	33.9	3.3	67.4	58.8	87.3	905	912	1.01	145
Epox10	0.313	12.3	1.8	88.2	76.3	86.6	908	915	1.01	218
Epox15	0.334	5.8	2.3	94.4	81.5	86.3	914	921	1.01	226
Epox20	0.339	2.6	2.9	97.5	82.7	84.9	919	926	1.01	266

Ev—content of epoxy groups; Iv—iodine value; Hv—hydroxyl value; E—efficiency; C—conversion; S—selectivity; Mn—number average molecular weight; Mw—weight average molecular weight; D—dispersity; η—viscosity.

As the concentration of the catalyst used in the reaction increases, the viscosity of the epoxidized oils is increased. S. Zoran et al. have shown that an increase in the degree of conversion to oxirane rings correlates with an increase in the viscosity of an epoxidized oil, which is a result of side reactions that can also increase viscosity [13]. There are no major changes in the hydroxyl or acid value that can be correlated with the change in the concentration of the catalyst used in the reaction.

Dinda et al. analyzed the influence of the AIER concentration (10–25 wt.%) on the epoxidation reaction of cottonseed oil. They concluded that when the catalyst loading was increased from 10 to 15 wt.%, the oxirane conversion increased and it was on the same level as after an addition of a greater amount of the catalyst [18]. In the case of Jatropha oil epoxidation, Goud at al. obtained similar results. The maximum conversion to oxirane in an epoxidation reaction conducted with 16% Amberlite IR-120 loading was lower by only 2.7% than that achieved with 20% [28]. It can be concluded that the process of epoxidation of waste oil is the same as in the case of fresh oils and does not require higher catalyst loading due to the presence of oxidation products in used cooking oil.

The efficiency (E), conversion (C), and selectivity (S) of all reactions (Table 1) were found according to the following equations:

$$E = \frac{Ev}{Ev_{max}} \cdot 100\% \tag{1}$$

Ev—the epoxy number of the ester of a vegetable oil after the epoxidation reaction, mol/100 g of epoxidized oil

Ev_{max}—the epoxy number calculated based on the number of unsaturated bonds, mol/100 g of epoxidized oil

$$C = \frac{Iv_0 - Iv}{Iv_0} \cdot 100\% \tag{2}$$

Iv_0—the iodine number of the methyl ester/vegetable oil before epoxidation, gI$_2$/100 g of oil.
Iv—the iodine number of the methyl ester/vegetable oil after epoxidation, gI$_2$/100 g of oil.

$$S = \frac{Ev}{Iv_0 - Iv} \cdot 100\% \tag{3}$$

In the reactions carried out, the maximum conversion rate was 97.5%. Espinoza Perez et al. have obtained a conversion rate of 98.5% for the rapeseed oil epoxidation reaction [29]. Espinoza Perez et al. showed a significant effect of conducting the reaction in a solvent (toluene) environment that allows higher conversion rates than the analogous reaction carried out without a solvent [29].

Oil epoxidation may affect its average molar mass. In the environment of epoxidation reactions, partial triglyceride breakdown may occur, resulting in lower molecular weight products such as mono- and diglycerides, free fatty acids, and glycerin. Newly formed epoxide rings may also be opened, and derivatives of higher molecular weight may be formed, which in turn increases the molecular weight. Both trends result in an increase in the dispersion of the resulting reaction product [5].

Based on our research, we concluded that the content of the heterogeneous catalyst did not have a significant effect on the average molar masses of epoxidized waste oils and the compounds obtained were monodisperse. The low dispersion value of the epoxidized oils obtained indicates a small amount of by-products resulting from reactions such as oligomers.

In order to find the effect of particle size and degree of cross-linking AIER on the properties of the epoxidized oil, the experiments were conducted using three different sizes of resin particles as well as resins containing three different amounts of the cross-linking agent. Nine different Dowex resins were used in the study and the results were compared with commonly used Amberlite IR-120. The characteristics of the heterogeneous catalysts are presented in the Table 2 and photographs of the catalysts are shown in Figure 3.

The progress of the reaction was monitored by determining the epoxy number and by a FTIR analysis during the reaction. The epoxy number changes during the reaction and the changes of characteristic bands in the FTIR spectrum are presented in Figures 4 and 5.

Table 2. Characteristics of acidic ion exchange resins (AIER).

AIER		Content of Cross-Linking Agent in Resin, %	Name of Oil Sample	Grain Size, mm	Moisture Content, %	Price [PLN/100 g]
Dowex 50WX2	50–100	2	Epox_2_50-100	0.15–0.30	93.2	713
	100–200	2	Epox_2_100-200	0.07–0.15	87.2	699
	200–400	2	Epox_2_200-400	0.04–0.07	84.9	699
Dowex 50WX4	50–100	4	Epox_4_50-100	0.15–0.30	89.3	600
	100–200	4	Epox_4_100-200	0.07–0.15	89.4	538
	200–400	4	Epox_4_200-400	0.04–0.07	87.9	500
Dowex 50WX4	50–100	8	Epox_8_50-100	0.15–0.30	89.9	523
	100–200	8	Epox_8_100-200	0.07–0.15	89.7	500
	200–400	8	Epox_8_200-400	0.04–0.07	93.5	476
Amberlite IR-120		8	Epox_15	0.62–0.83	85.0	32.1

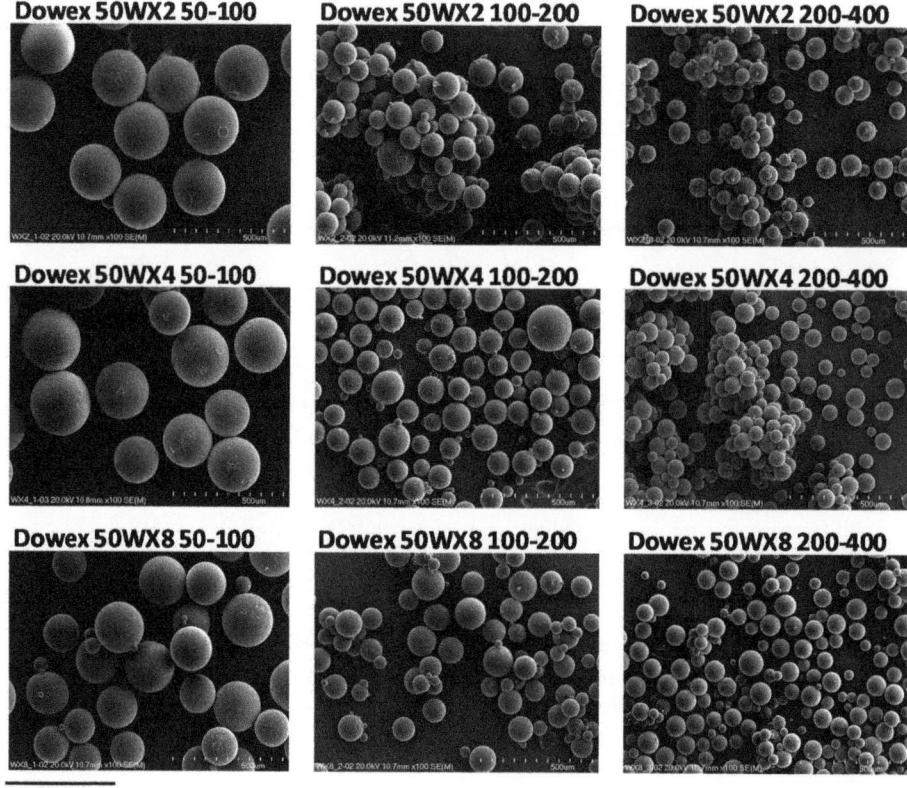

Figure 3. SEM micrographs of Dowex ion exchange resins.

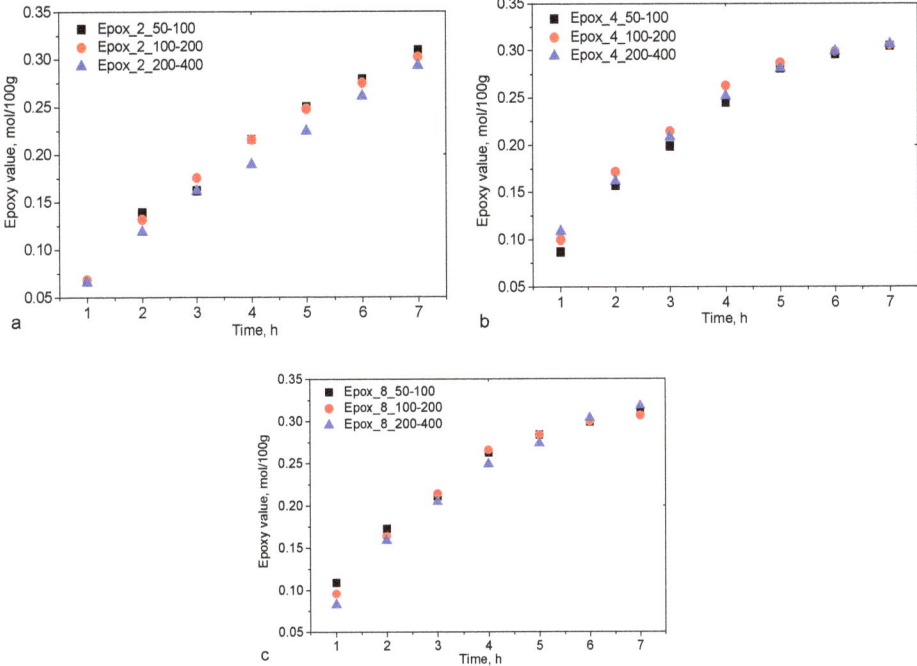

Figure 4. Influence of different grain size of (**a**) Dowex50 × 2, (**b**) Dowex50 × 4, and (**c**) Dowex50 × 8.

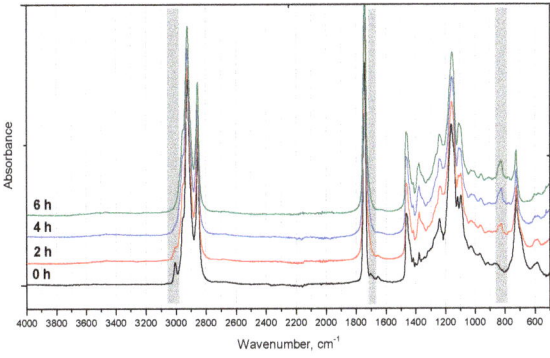

Figure 5. FTIR spectra of samples taken during epoxidation process of Epox_8_200-400.

During all processes, the Ev increased as the reaction progressed, which indicates double bond oxidation in fatty acid chains to form oxirane rings (Figure 4). Slightly lower initial epoxy values were noted for the reaction in the presence of a catalyst containing 2% of the cross-linker. However, no significant difference was observed in the final Evs depending on the type of the catalyst used in the synthesis.

The analysis of FTIR spectra collected during the reaction allowed on-line detection of characteristic groups. In Figure 5, the FTIR spectra of the used cooking oil epoxidation with Dowex50X8 200-400 are shown. In the FTIR spectrum of the waste rapeseed oil, with wave numbers of about 3010 cm^{-1} and about 1650 cm^{-1}, the bands corresponding to the vibrations of the double bonds between carbon atoms are visible. These signals lose their intensity as the epoxidation reaction progresses. Along with the decrease in the intensity of the bands corresponding to the vibrations of unsaturated bonds, the FTIR spectra of the epoxidized oil samples show signals in the range 800–850 cm^{-1}, which are characteristic for epoxy groups.

In all spectra, apart from the signals of epoxy groups, there are also bands at about 2925 and 2850 cm^{-1}, characterized by high intensity. They arise as a result of stretching vibrations of bonds between carbon and hydrogen atoms in the groups –CH$_2$– and –CH$_3$, occurring in fatty acid chains. Another characteristic band in the spectra of the epoxidized oils is the signal visible at approx. 1740 cm^{-1}, which comes from the stretching vibrations C=O of the ester group.

Ev, Iv, and Hv determinations were carried out for the epoxidized oils obtained (Table 3).

Table 3. Characteristics of epoxidized oil obtained using different AIER.

Sample	Ev, mol/100 g	Iv, gI$_2$/100 g	Hv, mgKOH/g	C	E	S
Used cooking oil	-	104.0	-	-	-	-
Epox_2_50-100	0.346	9.76	8.89	91	84	93
Epox_2_100-200	0.319	11.11	12.29	89	78	87
Epox_2_200-400	0.324	7.09	12.89	93	79	85
Epox_4_50-100	0.331	9.96	11.29	90	81	89
Epox_4_100-200	0.344	6.36	10.01	94	84	89
Epox_4_200-400	0.337	6.74	12.32	94	82	88
Epox_8_50-100	0.338	8.76	9.85	92	82	90
Epox_8_100-200	0.341	7.54	9.79	93	83	90
Epox_8_200-400	0.353	8.35	8.93	92	86	94

Based on our research it was found that in the case of Dowex resins containing 2% cross-linking agent the highest epoxy value was obtained for the resin with the largest grain size. In the case of resins containing 8% divinylbenzene, the opposite effect was observed. However, the differences in the epoxy numbers are insignificant. Therefore, the results obtained were compared with the epoxy oil (Epox_15) obtained through a reaction with Amberlite resin. This resin has a much lower price compared to Dowex resins. From the point of view of waste oil recycling, it is important to develop a low-cost process. The epoxy number of Epox_15 is slightly lower than the epoxy numbers obtained for the oils synthesized in the presence of Dowex resins containing 8% divinylbenzene, and comparable to resins with 4% cross-linking agent. Dinda et al. conducted epoxidation of cottonseed oil using two very different sizes of resin particles under otherwise similar conditions. The particle sizes were greater than 620 μm and smaller than 120 μm. They concluded that both particle sizes gave nearly the same oxirane conversion [18]. Goud et al. also analyzed the intraparticle diffusional limitations by using two widely differing particle sizes of resin, namely >599 and <64 μm under otherwise the same reaction conditions. They concluded that both particle sizes gave practically the same results [30].

During all epoxidation processes, double bond conversion in the range of 89–94% was achieved. The conversion value for the reaction with Amberlite was 94%. Comparing to the literature data and cases where fresh oils were epoxidized, the values obtained in this study are very favorable. According to literature reports, during oil epoxidation using acetic peroxyacid and in the presence of Amberite IR-120 ion exchange resin, it is possible to achieve about 90% efficiency after 7 h of the process at 65 °C [15]. The oxidation reaction of unsaturated soybean oil bonds, using the same oxidizing agent and catalyst for 10 h and at 60 °C allows achieving about 86.8% double bond conversion and about 86% selectivity [31]. On the other hand, the epoxidation of karanja oil with the use of acetic peracid and Amberlite IR-120 resin leads to approx. 85% yield.

Waste rapeseed oil and its derivatives, in the form of epoxidized oils, were also characterized by a similar molar mass distribution (Figure 6).

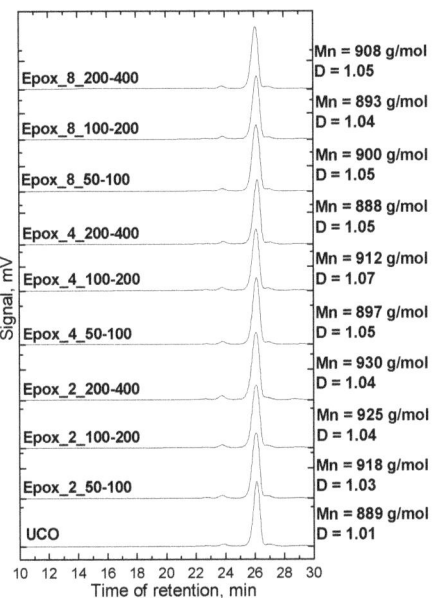

Figure 6. Chromatograms of bio-polyols obtained with Dowex catalysts.

The highest intensity signal present in all chromatograms corresponds to the weight of triglyceride. Its retention time is about 26 min. Peaks characterized by low intensity, with a retention time of about 24 min, indicate the presence of oligomers in the epoxidized oils. The signal is also visible in the chromatogram of waste rapeseed oil, hence the oligomerization products may have arisen as a result of a processes related to the thermal treatment of vegetable oil. The waste oil and epoxidized oils were characterized by similar average molar masses.

The dispersion of the products was slightly higher than the dispersion of the starting oil and ranged from 1.03 to 1.07. A small dispersion of the masses of the synthesized epoxidized oils is their advantage. In the production on an industrial scale, striving to obtain materials with low dispersion is sought as it has a positive effect on their performance and the possibility of further processing.

The advantage of heterogeneous catalysts should be easiness of separating them from products. Our experiments showed that in the case of the processes catalyzed by Dowex ion exchange resins, due to their small size, it was not possible to remove the catalyst from the epoxidized oil thoroughly using methods available on a laboratory scale. This resulted in turbid products. For the Amberlite IR-120 resin, this problem did not occur. The epoxidation process in the presence of Dowex ion exchange resins is not economical resulting from high prices of catalysts. Catalysts with smaller grain diameters and lower cross-linker content are generally more expensive. In contrast, the use of Amberlite resin allows obtaining epoxidized oils that can be used as plasticizers as well as intermediates for subsequent syntheses whereas the costs of waste oil modification are reduced.

The life of AIER is limited due to deactivation after a few cycles and the disposal of used catalyst is a major problem faced by industry [32]. In order to create a more environmentally friendly process, the possibility of repeated use of the same catalyst has been analyzed in the literature. Such an experiment was described for the epoxidation of fresh vegetable oils such as cotton seed oil [18], castor oil [33]. In the case of used cooking oil, there are oxidation products and other impurities confirmed by high peroxide and anisidine values, as well as the presence of polycyclic aromatic hydrocarbons and polychlorinated biphenyls [34]. In order to analyze such impurities and their influence on the efficiency of AIER, in this part of our work the same catalyst was used ten times for the epoxidation reactions. The microphotographs of Amberlite after each reaction are shown in Figure 7.

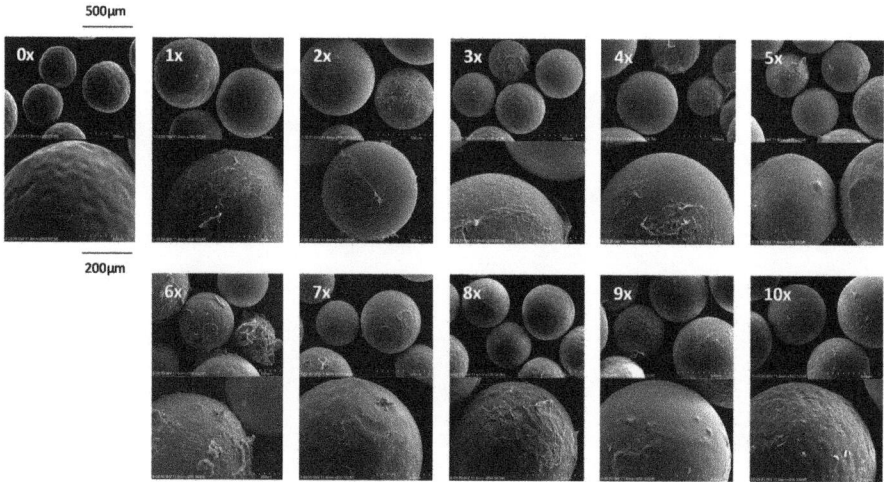

Figure 7. SEM microphotographs of Amberlite IR-120 catalyst before epoxidation reaction (0×) and after each reaction (1×–10×).

Regardless of the reaction, contaminants are present on the resin surface. This effect is associated with the method of resin preparation for synthesis. In order to minimize the costs and make the process simpler, we only separated the catalyst by filtration from the epoxidized oil and washed with water in order to remove acetic acid.

It was observed that after the first recycle the epoxy value of sample 2× was 6% lower (Figure 8). However, interesting results not described in the literature before were obtained in the case of the 3× and 4× reactions where the same catalyst was used for the 3rd and 4th time. The epoxide number of the modified oils was 8 and 6% higher, respectively, compared to the 1× reaction. This effect can be associated with the swelling of the catalyst, which results in easier access to the acid sulfonic groups of the resin.

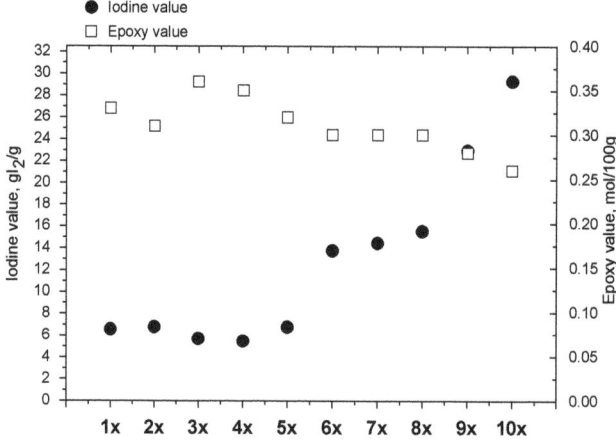

Figure 8. Changes in Ev and Iv depending on the number of times the catalyst was used.

In the literature there are descriptions of solid catalyst recycling. Dinda et al. used the same Amberlite IR-120 in four consecutive experiments and concluded that after the first recycle the catalyst activity decreased gradually with the number of recycles [18]. Mungroo et al. [17] found that AIER can be reusable and exhibited a negligible activity loss. After the catalyst had been used four consecutive times, the relative conversion to oxirane and the iodine conversion were 83 and 85%, as compared to 90 and 88.4%, respectively, obtained with the fresh catalyst. Goud et al. [16] based on their studies on the epoxidation of karanja oil found that it is possible to repeat recirculation of the catalyst Amberlite IR-120 four times. After each synthesis, the catalyst was regenerated. The regeneration relied on the filtration of resin, rinsing with water and diethyl ether, and drying in ambient temperature.

3. Materials and Methods

3.1. Materials

Glacial acetic acid (min. 99.5–99.9 wt.%), hydrogen peroxide (30 wt.%), were purchased from Avantor Performance Materials Poland S.A (Gliwice, Poland). Ion-exchange resins—Amberlite® IR-120 (Sigma-Aldrich, St. Louis, MO, USA), Dowex 50WX2 50-100, Dowex 50WX2 100-200, Dowex 50WX2 200-400, Dowex 50WX4 50-100, Dowex 50WX4 100-200, Dowex 50WX4 200-400, Dowex 50WX8 50-100, Dowex 50WX8 100-200, Dowex 50WX8 200-400 were purchased from Sigma-Aldrich (St. Louis, MO, USA). Used cooking oil was collected from 3 local restaurants (Kraków, Poland). The iodine and acid values of the used cooking oil mixtures were 104 gI_2/g.

3.2. Epoxidation Procedure

In the experiment, 250 g of used cooking oil and 0.5 mol of acetic acid and 2 mol hydrogen peroxide per mole of unsaturated bonds of oil as well ion exchange resin were added to a reactor. The content of Amberlite 120 was 15 wt.% (with respect to oil mass). The reactions were carried out for 6 h at a temperature of 60–65 °C using continuous stirring. Samples were taken out at the 1st, 2nd, 3rd, 4th, 5th and 6th h. After six hours the reaction mixture was separated into two phases, organic and aqueous. The organic phase was washed successively with warm water until it was acid free. The organic phase was distilled under vacuum (10 mbar) for 2 h in order to remove of water [5].

3.3. Methodology of Epoxidized Oil Characterization

The designation of Iv was done using the Hanus method according to the standard PN-87/C-04281, in which the iodine atoms are added to unsaturated bonds. The unsaturation degree of a given fat is then expressed by the amount of the iodine added.

Ev was determined according to the PN-87/C-89085/13 standard. The method involves a quantitative reaction of hydrogen chloride with a reactive epoxy group in dioxane at room temperature and titration of the hydrogen chloride excess using a solution of sodium hydroxide in methanol in the presence of cresol red as an indicator.

Hv of the polyol was determined according to the standard PN-93/C-89052/03, in which the hydroxy groups of a polyol undergo acetylation using acetic anhydride. The excess of the acetic anhydride is decomposed by a water addition (formation of acetic acid) and followed by titration using a solution of potassium hydroxide in the presence of an indicator.

Viscosity (η) was determined using a rotational rheometer HAAKE MARS III (Thermo Scientific, Waltham, MA, USA) at 25 °C. The control rate mode was used in the plate-plate arrangement with the plates having a diameter of 20 mm and rotation speeds of 100 cycles/min.

Number average molecular weight (M_n) and dispersity (Đ) were determined by a gel permeation chromatography (GPC) analysis. GPC measurements were performed using a Knauer chromatograph (Warsaw, Poland). The calibration was performed using polystyrene standards. Tetrahydrofuran (Avantor Performance Materials Poland S.A, Gliwice, Poland) was used as an eluent at a 0.8 mL/min flow rate at room temperature.

FTIR spectroscopy was performed using a FT-IR SPECTRUM 65 spectrometer (Perkin Elmer, Waltham, MA, USA).

4. Conclusions

The results presented in this paper allow verification of the influence of the heterogeneous catalyst concentration, its structure and particle size, as well as repeated use on the waste oil epoxidation process. No significant differences were observed in the final properties of the epoxidized waste cooking oil depending on the particle size and the cross-linking degree of the ion exchange resins used in the synthesis. The results were compared with the reaction in which Amberlite 120 resin was used, characterized by a larger grain size and a lower price. It was found that the Amberlite 120 resin can be successfully used in used cooking oil epoxidation reactions. Aiming at waste reduction in technological processes, an attempt was made to reuse a heterogeneous catalyst. The catalyst Amberlite IR-120 was reused without further treatment and no significant differences in the epoxy value were observed. Application of the same resin eight times in epoxidation allows obtaining products with epoxy values of 0.3 mol/100 g and higher. It was noticed that in the case of the reactions where the catalyst was reused for the third and fourth time the content of oxirane rings was higher by 8 and 6% compared to reaction where the catalyst was used one time only. Such an observation has not been reported so far.

Author Contributions: Conceptualization, M.K.; methodology, M.K.; validation, M.K.; formal analysis, M.K.; investigation, M.K. and M.N.; resources, M.K.; data curation, M.K. and M.N.; writing—original draft preparation, M.K.; writing—review and editing, M.K.; visualization, M.K.; supervision, M.K.; project administration, M.K.; funding acquisition, M.K. All authors have read and agreed to the published version of the manuscript.

Funding: This research was funded by National Center for Research and Development in Poland under the Lider Program, grant number LIDER/28/0167/L-8/16/NCBR/2017.

Conflicts of Interest: The authors declare no conflict of interest.

References

1. Saurabh, T.; Patnaik, M.; Bhagt, S.L.; Renge, V.C. Epoxidation of vegetable oils: A review. *Int. J. Adv. Eng. Technol.* **2011**, *4*, 491.
2. Guner, F.S.; Yagci, Y.; Erciyes, T. Polymers from triglyceride oils. *Prog. Polym. Sci.* **2006**, *31*, 633. [CrossRef]
3. Panadare, D.C.; Rathod, V.K. Applications of Waste Cooking Oil Other Than Biodiesel: A Review. *Iran. J. Chem. Eng.* **2015**, *12*, 55–76.
4. Kurańska, M.; Benes, H.; Polaczek, K.; Trhlikova, O.; Walterova, Z.; Prociak, A. Effect of homogeneous catalysts on ring opening reactions of epoxidized cooking oils. *J. Clean. Prod.* **2019**, *230*, 162. [CrossRef]
5. Kurańska, M.; Benes, H.; Prociak, A.; Trhlíkova, O.; Walterova, Z.; Stochlińska, W. Investigation of epoxidation of used cooking oils with homogeneous and heterogeneous catalysts. *J. Clean. Prod.* **2019**, *236*, 117615. [CrossRef]
6. Turco, R.; Di Serio, M. Sustainable Synthesis of Epoxidized Cynara, C. Seed Oil. *Catalyst* **2020**, *10*, 721. [CrossRef]
7. Dinda, S.; Patwardhan, A.V.; Goud, V.V.; Pradhan, N.C. Epoxidation of cottonseed oil by aqueous hydrogen peroxide catalysed by liquid inorganic acids. *Bioresour. Technol.* **2008**, *99*, 3737. [CrossRef]
8. Milchert, E.; Smagowicz, A. The Influence of Reaction Parameters on the Epoxidation of Rapeseed Oil with Peracetic Acid. *J. Am. Oil Chem. Soc.* **2009**, *86*, 1227. [CrossRef]
9. Milchert, E.; Malarczyk-Matusiak, K.; Musik, M. Technological aspects of vegetable oils epoxidation in the presence of ion exchange resins: A review. *Pol. J. Chem. Technol.* **2016**, *18*, 128. [CrossRef]
10. Patil, H.; Waghmare, J. Catalyst for epoxidation of oils: A review. *Discovery* **2013**, *3*, 10–14.
11. Goud, V.V.; Patwardhan, A.V.; Pradhan, N.C. Studies on the epoxidation of mahua oil (Madhumica indica) by hydrogen peroxide. *Bioresour. Technol.* **2006**, *97*, 1365. [CrossRef] [PubMed]
12. Silviana, S.; Anggoro, D.D.; Kumoro, A.C. Waste Cooking Oil Utilisation as Bio-plasticiser through Epoxidation using Inorganic Acids as Homogeneous Catalysts. *Chem. Eng. Trans.* **2017**, *56*, 1861.
13. Petrović, Z.S.; Zlatanić, A.; Lava, C.C.; Sinadinović-Fišer, S. Epoxidation of soybean oil in toluene with peroxoacetic and peroxoformic acids-Kinetics and side reactions. *Eur. J. Lipid Sci. Technol.* **2002**, *104*, 293–299. [CrossRef]
14. Rios, L.A.; Echeverri, D.A.; Franco, A. Epoxidation of jatropha oil using heterogeneous catalysts suitable for the Prileschajew reaction: Acidic resins and immobilized lipase. *Appl. Catal. A Gen.* **2011**, *394*, 132. [CrossRef]

15. Mungroo, R.; Goud, V.V.; Naik, S.N.; Dalai, A.K.; Reaction, C.; Laboratories, E.; Delhi, N. Utilization of green seed canola oil for in situ epoxidation. *Eur. J. Lipid Sci. Technol.* **2011**, *113*, 768. [CrossRef]
16. Goud, V.V.; Patwardhan, A.V.; Dinda, S.; Pradhan, N.C. Epoxidation of karanja (Pongamia glabra) oil. *Eur. J. Lipid Sci. Technol.* **2007**, *109*, 575. [CrossRef]
17. Mungroo, R.; Pradhan, N.C.; Goud, V.V.; Dalai, A.K. Epoxidation of Canola Oil with Hydrogen Peroxide Catalyzed by Acidic Ion Exchange Resin. *J. Am. Oil Chem. Soc.* **2008**, *85*, 887. [CrossRef]
18. Dinda, S.; Goud, V.V.; Patwardhan, A.V.; Pradhan, N.C. Selective epoxidation of natural triglycerides using acidic. *Asia-Pac. J. Chem. Eng.* **2011**, *6*, 870. [CrossRef]
19. Goud, V.V.; Pradhan, N.C.; Patwardhan, A.V. Epoxidation of Karanja (Pongamia glabra) Oil by H_2O_2. *J. Am. Oil Chem. Soc.* **2006**, *83*, 635. [CrossRef]
20. Narowska, B.E.; Kułażyński, M.; Łukaszewicz, M. Application of Activated Carbon to Obtain Biodiesel from Vegetable Oils. *Catalysts* **2020**, *10*, 1049. [CrossRef]
21. Vlcek, T.; Petrovic, Z.S. Optimization of the Chemoenzymatic Epoxidation of Soybean Oil. *J. Am. Oil Chem. Soc.* **2006**, *83*, 247. [CrossRef]
22. Rusch, M.; Warwel, S. Complete and partial epoxidation of plant oils by lipase-catalyzed perhydrolysis. *Ind. Crop. Prod.* **1999**, *9*, 125. [CrossRef]
23. Sun, S.; Ke, X.; Cui, L.; Yang, G.; Bi, Y.; Song, F.; Xu, X. Enzymatic epoxidation of Sapindus mukorossi seed oil by perstearic acid optimized using response surface methodology. *Ind. Crop. Prod.* **2011**, *33*, 676. [CrossRef]
24. Mindaryani, A.; Rahayu, S.S. Epoxidation of Candlenut Oil. In Proceedings of the 2010 International Conference on Chemistry and Chemical Engineering, Kyoto, Japan, 1–3 August 2010; p. 102.
25. Goud, V.V.; Dinda, S.; Patwardhan, A.V.; Pradhan, N.C. Epoxidation of Jatropha (Jatropha curcas) oil by peroxyacids. *Asia-Pac. J. Chem. Eng.* **2010**, *5*, 346. [CrossRef]
26. Monono, E.M.; Haagenson, D.M.; Wiesenborn, D.P. Characterizing the epoxidation process conditions of canola oil for reactor scale-up. *Ind. Crop. Prod.* **2015**, *67*, 364. [CrossRef]
27. Chieng, B.W.; Ibrahim, N.A.; Then, Y.Y.; Loo, Y.Y. Epoxidized Vegetable Oils Plasticized Poly(lactic acid) Biocomposites: Mechanical, Thermal and Morphology Properties. *Molecules* **2014**, *19*, 16024. [CrossRef]
28. Goud, V.V.; Patwardhan, A.V.; Dinda, S.; Pradhan, N.C. Kinetics of epoxidation of jatropha oil with peroxyacetic and peroxyformic acid catalysed by acidic ion exchange resin. *Chem. Eng. Sci.* **2007**, *62*, 4065. [CrossRef]
29. Perez, J.D.; Haagenson, D.M.; Pryor, S.W.; Ulven, C.A.; Wiesenborn, D.P. Production and characterization of epoxidized canola oil. *ASABE* **2009**, *52*, 1289. [CrossRef]
30. Goud, V.V.; Patwardhan, A.V.; Pradhan, N.C. Kinetics of in situ Epoxidation of Natural Unsaturated Triglycerides Catalyzed by Acidic Ion Exchange Resin. *Ind. Eng. Chem. Res.* **2007**, *46*, 3078. [CrossRef]
31. Sinadinović-Fišer, S.; Janković, M.; Petovic, Z.S. Kinetics of in situ Epoxidation of Soybean Oil in Bulk Catalyzed by Ion Exchange Resin. *J. Am. Oil Chem. Soc.* **2001**, *78*, 725. [CrossRef]
32. Malshe, V.C.; Sujatha, E.S. Regeneration and reuse of cation-exchange resin catalyst used in alkylation of phenol. *React. Funct. Polym.* **1997**, *35*, 159. [CrossRef]
33. Sinadinović-Fišer, S.; Janković, M.; Borota, O. Epoxidation of castor oil with peracetic acid formed in situ in the presence of an ion exchange resin. *Chem. Eng. Process. Process. Intensif.* **2012**, *62*, 106. [CrossRef]
34. Kurańska, M.; Banaś, J.; Polaczek, K.; Banaś, M.; Prociak, A.; Kuc, J.; Uram, K.; Lubera, T.J. Evaluation of application potential of used cooking oils in the synthesis of polyol compounds. *Environ. Chem. Eng.* **2019**, *7*, 103506. [CrossRef]

Publisher's Note: MDPI stays neutral with regard to jurisdictional claims in published maps and institutional affiliations.

© 2020 by the authors. Licensee MDPI, Basel, Switzerland. This article is an open access article distributed under the terms and conditions of the Creative Commons Attribution (CC BY) license (http://creativecommons.org/licenses/by/4.0/).

Article

Fe₃O₄-Zeolite Hybrid Material as Hetero-Fenton Catalyst for Enhanced Degradation of Aqueous Ofloxacin Solution

Alamri Rahmah Dhahawi Ahmad, Saifullahi Shehu Imam, Wen Da Oh and Rohana Adnan *

School of Chemical Sciences, Universiti Sains Malaysia, Penang 11800, Malaysia; ralamri-1406@outlook.com (A.R.D.A.); ssimam.chm@buk.edu.ng (S.S.I.); ohwenda@usm.my (W.D.O.)
* Correspondence: r_adnan@usm.my

Received: 11 September 2020; Accepted: 22 October 2020; Published: 27 October 2020

Abstract: A hetero-Fenton catalyst comprising of Fe_3O_4 nanoparticles loaded on zeolite (FeZ) has been synthesized using a facile co-precipitation method. The catalyst was characterized using various characterization methods and then, subsequently, was used to degrade ofloxacin (OFL, 20 mg·L^{-1}), an antibiotic, via a heterogeneous Fenton process in the presence of an oxidizing agent. The effects of different parameters such as Fe_3O_4 loading on zeolite, catalyst loading, initial solution pH, initial OFL concentration, different oxidants, H_2O_2 dosage, reaction temperature, and inorganic salts were studied to determine the performance of the FeZ catalyst towards Fenton degradation of OFL under different conditions. Experimental results revealed that as much as 88% OFL and 51.2% total organic carbon (TOC) could be removed in 120 min using the FeZ catalyst. Moreover, the FeZ composite catalyst showed good stability for Fenton degradation of OFL even after five cycles, indicating that the FeZ catalyst could be a good candidate for wastewater remediation.

Keywords: Fenton degradation; ofloxacin; Fe_3O_4; zeolite; heterogeneous

1. Introduction

Ofloxacin (OFL) (Figure 1) is a second-generation fluoroquinolone antibiotic with the chemical formula $C_{18}H_{20}FN_3O_4$ and the chemical name 9-fluoro-2,3-dihydro-3-methyl-10-(4-methyl-1-piperazynyl)-7-oxo-7H-pyrido-[1,2,3-de]-1,4-benzoxazine-6-carboxylic acid [1,2]. It was patented in 1980 and subsequently approved for medical use in 1985 [3,4]. Currently, OFL is frequently prescribed for the treatment of various bacterial infections that cause digestive, respiratory, gastrointestinal, and urinary tract infections [5,6]. However, due to its partial metabolism in the body after ingestion, biological resistance, and the large volume of pharmaceutical wastewater which has been released untreated, studies have reported the detection of OFL with different concentrations in hospital wastewater (25,000–35,000 ng·L^{-1}), municipal wastewater treatment plants (53–1800 ng·L^{-1}), and surface water (10–535 ng·L^{-1}), with a residence time of about 10.6 days [7–9]. The presence of OFL in water results in unpleasant odors [10]. Besides, it may also lead to microbial resistance among pathogens or the death of microorganisms that are effective in wastewater remediation [7,10]. Although, currently, pharmaceutical compounds, including OFL, belong to the emerging pollutants category that is still not regulated by water quality laws, the identification of proper process(es) for their complete elimination from wastewater is imperative [11].

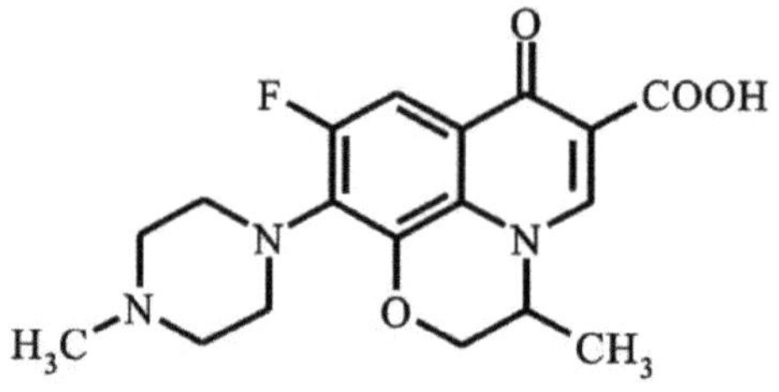

Figure 1. Chemical structure of ofloxacin (OFL).

Unfortunately, the techniques employed by most wastewater treatment plants (WWTPs) have limited capacity for the thorough elimination of pharmaceuticals and personal care products (PPCPs) from wastewater, as they are not originally designed for removing PPCPs [12]. Furthermore, certain physical and biological methods that are also being employed have some limitations. For instance, most of the physical treatment methods only transfer the pollutants to another phase rather than destroying them [12]. Likewise, a biological treatment method, such as biodegradation, could lead to the development of antibiotic-resistant bacteria [13].

Unlike the physical and biological treatment methods, chemical oxidation methods such as homogeneous and heterogeneous Fenton treatment methods are capable of mineralizing a wide range of organic pollutants [14]. However, compared to the homogeneous Fenton process, the easy recovery of the catalyst after application in the case of a heterogeneous Fenton process makes it more convenient [15]. Following that, many researchers have reported various methods of enhancing the efficiency of a heterogeneous Fenton process. The common method involves immobilizing the metallic ions/oxides onto various supports. Such supports, including activated carbon [16], carbon nanotubes [17], graphite oxide [18], SBA-15 [19], etc., are capable of improving the efficiency of the Fenton process, and also ease the catalyst recovery process after application. In the current work, Fe_3O_4-zeolite composites will be synthesized via a co-precipitation method for subsequent use as a catalyst in a heterogeneous Fenton process. To the best of our knowledge, there have been no studies on the heterogeneous Fenton degradation of OFL using zeolite-supported magnetite (Fe_3O_4-zeolite). This work reports the facile production of Fe_3O_4-zeolite as a low-cost catalyst for the heterogeneous Fenton degradation using a widely available and abundant material. In addition, zeolite has been used previously for the adsorption of organic pollutants [20,21]. The ability of zeolite to behave as an adsorbent implies a greater possibility of contact between catalyst and pollutant. Such a process will improve the efficiency of the Fenton reaction.

2. Results and Discussion

2.1. X-ray Photoelectron Spectroscopy (XPS) Analysis

The elemental composition and the interaction between Fe_3O_4 and zeolite in FeZ composites were studied using XPS analysis in the region of 0–1200 eV, and the results are shown in Figure 2. Based on the survey spectra presented in Figure 2a, C, O, Na, Al, and Si are present on the surface of bare zeolite, while the survey spectra in Figure 2b–f confirmed the coexistence of C, O, Al, Si, and Fe in the FeZ composites.

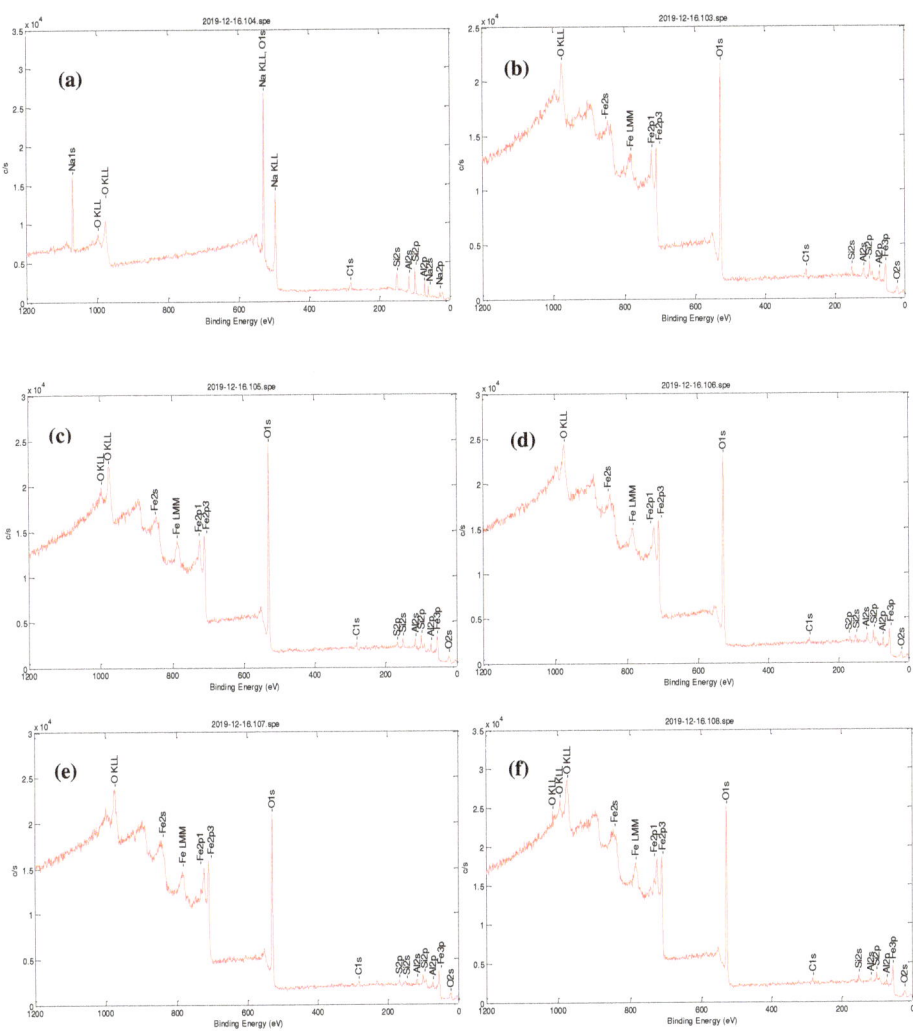

Figure 2. XPS spectra of (**a**) bare zeolite, (**b**) FeZ-1.5, (**c**) FeZ-3, (**d**) FeZ-5, (**e**) FeZ-8, and (**f**) FeZ-10 composites.

Within the FeZ composites spectra, two peaks with the binding energy of 710 and 725 eV are attributed to Fe $2p_{3/2}$ and Fe $2p_{1/2}$, indicating the presence of Fe_3O_4 [22]. Meanwhile, the peak around 95 eV also indicates the existence of Fe_3O_4 [23]. These results have proved the existence of iron oxide in the form of Fe_3O_4 in FeZ composites. The O 1s peak at 530 eV belongs to O^{2-} [24]. The presence of a peak centered at 286 eV is attributed to the presence of elemental carbon (C 1s) from the carbon band used during sample preparation. No other impurities were detected.

Finally, the XPS analysis result revealed that there is no Na^+ ion in the survey spectra of the FeZ composites. Such an effect is attributed to the exchange of sodium ions with iron ions in the interlayer of zeolite [25]. Based on the XPS analysis results, it is believed that composites of zeolite loaded with variable amounts of Fe_3O_4 have been successfully synthesized.

2.2. Fourier Transform Infrared (FTIR) Analysis

The functional groups present in bare zeolite and in FeZ composites were also studied using FTIR spectrometer, and the results are shown in Figure 3. In the case of bare zeolite, the absorption bands at 1020 and 440 cm^{-1} are due to Si–O–Si stretching and bending vibrations, respectively. Meanwhile, the absorption band at 532 cm^{-1} is due to Al–O–Si deformation [26]. However, the intensities of the peaks at 440 and 532 cm^{-1} attributed to Si–O–Si and Al–O–Si decrease significantly due to the loading of zeolite with Fe_3O_4. The peaks at 3500 and 1650 cm^{-1} are due to the stretching and bending vibrations of hydroxyl groups of absorbed water molecules, respectively [25]. In the case of FeZ composites, the band at 583 cm^{-1} is attributed to the stretching vibration of Fe–O [27], which confirmed the existence of Fe_3O_4 in FeZ composites [28]. The absence of such band in FeZ-1.5 and FeZ-3 composites might be attributed to the low amount of magnetite in the composites. Furthermore, the small peak at around 1450 cm^{-1} in the FeZ composites is due to the remnant of NH_3 used during composites preparation [25].

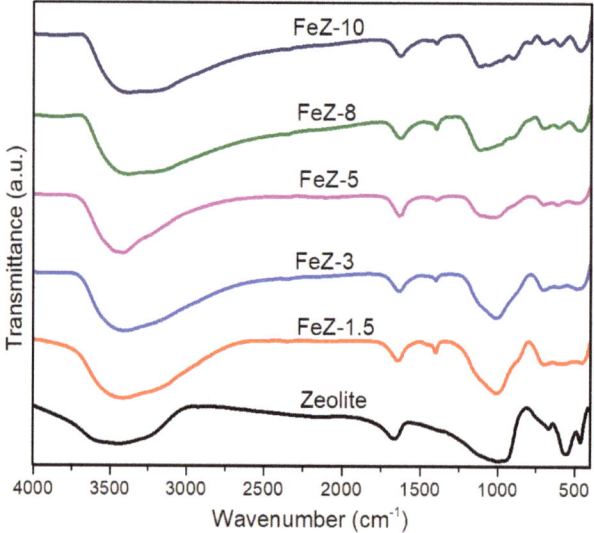

Figure 3. FTIR spectra of bare zeolite and FeZ composites.

2.3. Field Emission Scanning Electron Microscopy (FESEM)/Energy Dispersive X-ray (EDX) Analysis

FESEM images of bare zeolite and FeZ composites are shown in Figure 4. Based on the image presented in Figure 4a, bare zeolite is made up of various sized particles with an irregular shape, and the surface is virtually smooth. However, in the case of FeZ composites in Figure 4b–f, the surface is rough, and this could be due to the presence of Fe_3O_4 particles in the form of clusters or nanoparticles on the surface of the zeolite. Among the FeZ composites, the zeolite matrix is still clearly observable in the case of FeZ-1.5, FeZ-3, and FeZ-5 composites, which contain a low amount of Fe_3O_4 loading during synthesis. However, at a higher amount of Fe_3O_4, such as FeZ-8 and FeZ-10 composites, the presence of zeolite matrix cannot be detected.

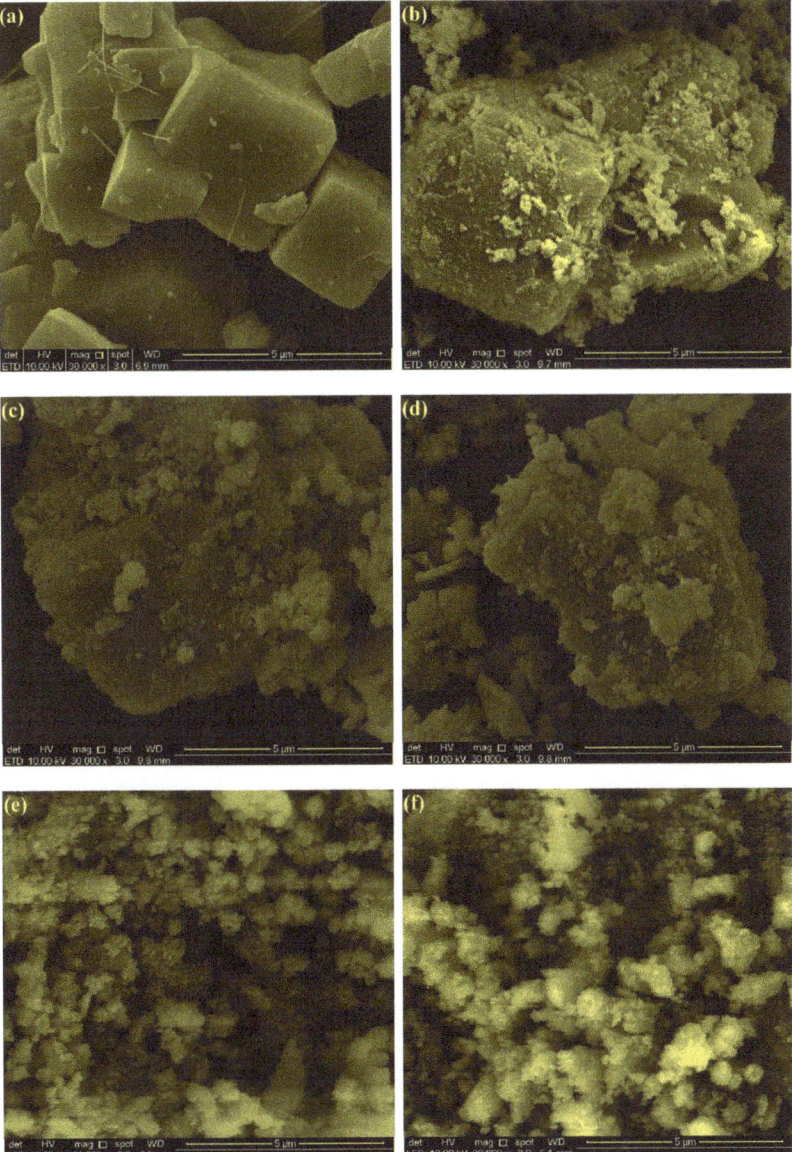

Figure 4. FESEM images of zeolite (**a**), FeZ-1.5 (**b**), FeZ-3 (**c**), FeZ-5 (**d**), FeZ-8 (**e**), and FeZ-10 (**f**) composites.

Furthermore, the elemental composition of bare zeolite and FeZ composites obtained from EDS analysis is shown in Table 1. The results revealed the presence of Na, Al, O, and Si in bare zeolite and Fe, Al, O, and Si in FeZ composites. The absence of Na in FeZ composites could be explained due to the ion exchange reaction with Fe ions. Moreover, the concentration of Fe in the FeZ composites keeps on increasing with the increase in the amount of loaded magnetite. Such results are a further indication that Fe_3O_4 has been immobilized onto zeolite.

Table 1. Composition of bare zeolite and FeZ composites as determined by EDX.

Sample	Al%	Si%	Na%	O%	Fe%
Zeolite	16.76	18.69	14.40	50.15	-
FeZ-1.5	17.81	17.84	-	41.67	22.68
FeZ-3	15.20	15.77	-	40.00	29.03
FeZ-5	15.02	13.91	-	39.47	31.59
FeZ-8	4.12	4.39	-	38.88	52.62
FeZ-10	4.09	4.01	-	35.60	56.29

2.4. Transmission Electron Microscopy (TEM) Analysis

The morphology of pristine zeolite and FeZ composites was further studied using TEM, and the results are shown in Figure 5. Based on the image presented in Figure 5a, pristine zeolite is composed of irregularly shaped particles with a smooth surface. In the case of FeZ composites shown in Figure 5b–f, some new particles apart from that of pristine zeolite were observed in the composites. The amount of the particles increases with the increase in the amount of Fe_3O_4 nanoparticles being immobilized onto zeolite. The change in morphology to hair-like particles was observed at higher magnetite concentration. In general, despite subjecting the FeZ composites to strong ultrasonication treatment during TEM sample preparation, the Fe_3O_4 nanoparticles could still be found on the zeolite surface. This indicates a strong interaction between Fe_3O_4 nanoparticles and zeolite.

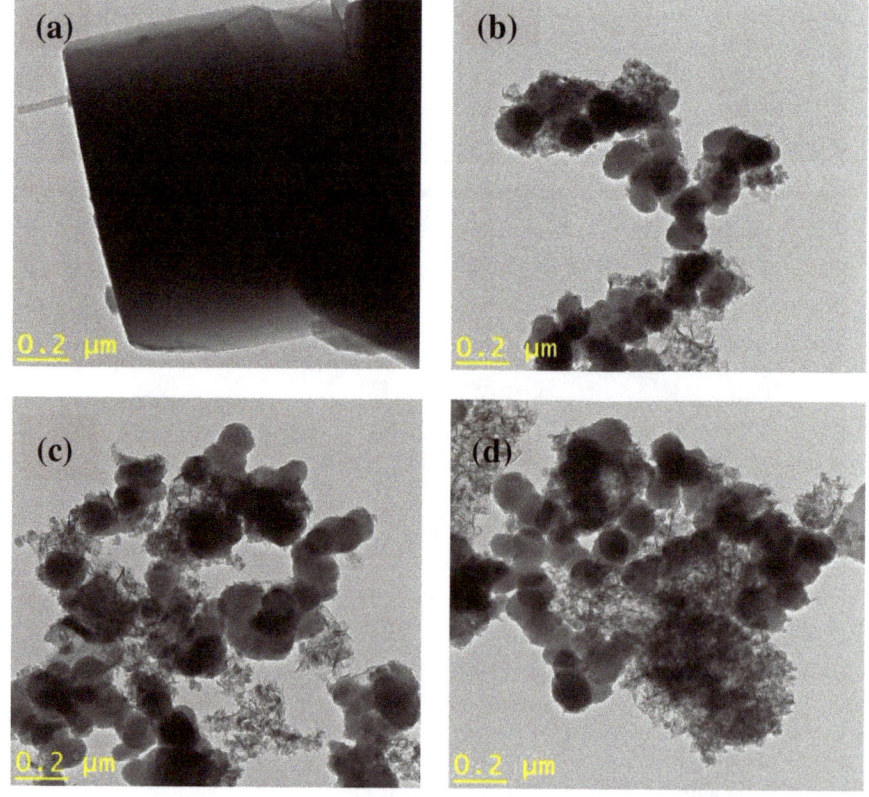

Figure 5. *Cont.*

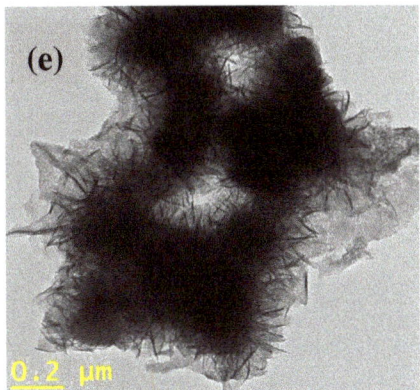 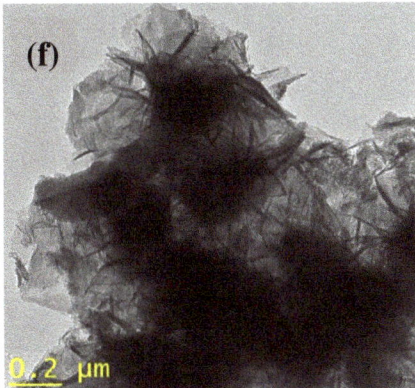

Figure 5. TEM images of zeolite (**a**), FeZ-1.5 (**b**), FeZ-3 (**c**), FeZ-5 (**d**), FeZ-8 (**e**), and FeZ-10 (**f**) composites.

2.5. Nitrogen Adsorption-Desorption Analysis

Nitrogen adsorption–desorption isotherms and Barrett-Joyner-Halender (BJH) pore size distribution (insets) of bare zeolite and FeZ composites are shown in Figure S1, and their structural details are summarized in Table 2. Although both bare zeolite and FeZ composites have type IV isotherms, FeZ composites have a distinct hysteresis loop in the range of P/P_0 0.4–1.0, due to capillary condensation of mesopores between closely packed spherical particles [29]. However, the adsorption capacities of the FeZ composites for N_2 are higher than that of bare zeolite and increase by as much as 160-fold as compared to zeolite alone. In addition, the shapes of the adsorption-desorption isotherms and pore size distributions of FeZ composites are different from that of bare zeolite, indicating the change in the pore structure due to the presence of Fe_3O_4 [30]. Furthermore, from the data presented in Table 2, the bare zeolite used has a low surface area (1.52 $m^2 \cdot g^{-1}$). However, the surface area of FeZ composites increased up to 251.6 $m^2 \cdot g^{-1}$ and decreases slightly to 211 $m^2 \cdot g^{-1}$ for FeZ-10. The decrease could be attributed to the agglomeration of Fe_3O_4 particles at higher loading. From the pore size distribution curves (insets), zeolite is polymodal. However, loading of Fe_3O_4 onto zeolite leads to the decrease in the pore size, an effect attributed to the blocking of zeolite pores by the magnetite. Meanwhile, the pore size distribution curve of FeZ-10 is bimodal, and this can be explained due to the formation of pores between the excess magnetite.

Table 2. Surface area and pore volume of bare zeolite and FeZ composites.

Samples	BET Surface Area ($m^2 \cdot g^{-1}$)	Pore Volume ($cm^3 \cdot g^{-1}$)
Zeolite	1.52	0.001798
FeZ-1.5	70.91	0.202736
FeZ-3	251.60	0.317371
FeZ-5	251.53	0.276022
FeZ-8	220.52	0.273748
FeZ-10	211.96	0.262692

2.6. Fenton Oxidation Activity of FeZ Composites

2.6.1. OFL Removal in Different Processes

Initially, a series of control experiments were conducted to determine the performance of (i) raw zeolite + OFL, (ii) H_2O_2 + OFL, (iii) Fe_3O_4-zeolite + OFL, (iv) zeolite + H_2O_2, and (v) Fe_3O_4-zeolite + H_2O_2 + OFL towards the removal of OFL. The Fe_3O_4-zeolite composite used for this study is FeZ-8, and the results are shown in Figure 6.

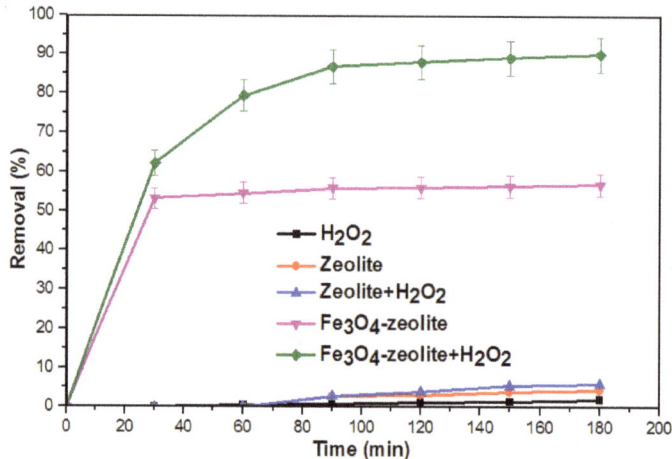

Figure 6. Removal efficiencies of OFL in different processes.

As shown in Figure 6, only 2% degradation was achieved after 180 min in the presence of H_2O_2 alone, which indicates that degradation under H_2O_2 is minimal. The results of such nature could be due to the low oxidation potential of H_2O_2 as compared to its $HO^•$ [31]. In the case of bare zeolite, the low percentage of removal of OFL (4%) achieved is attributed to its adsorption capability. The addition of H_2O_2 onto bare zeolite did not improve the efficiency, since zeolite is not capable of dissociating H_2O_2 to produce $HO^•$. Previously, Hu, et al. [32] also reported similar results. Meanwhile, Fe_3O_4-zeolite was able to adsorb up to 56.9% of OFL from the aqueous solution. This indicates that loading Fe_3O_4 onto bare zeolite leads to the formation of composites with higher adsorption capability towards OFL. On the other hand, in the presence of H_2O_2, the performance of Fe_3O_4-zeolite increased significantly to about 90%, thus indicating the reaction occurs via Fenton oxidation. Such results imply that Fe_3O_4-zeolite exhibited an excellent catalytic ability to activate H_2O_2, which is useful for Fenton reaction [33]. The results also confirmed the synergetic effect between Fe_3O_4-zeolite and H_2O_2 is necessary to achieve higher efficiency, and that dissociation of H_2O_2 is mainly caused by Fe_3O_4. The results obtained for the Fenton degradation of OFL using Fe_3O_4-zeolite composite was compared with the previous results reported for the Fenton degradation of OFL using different catalysts, and the results are shown in Table 3. The performance recorded using the Fe_3O_4-zeolite composite is encouraging and therefore, suggests that FeZ is a potential candidate for the treatment of wastewater contaminated by OFL.

Table 3. Comparison of Fenton catalytic performance in the removal of OFL.

Samples	Conc. of OFL (mg·L^{-1})	Catalyst Dosage (g·L^{-1})	Time (min)	pH	H$_2$O$_2$ Conc.	% Removal	Ref.
D-FeCu@Sep	10	3.0	120	5.0	0.03 M	~100	[34]
A-FeCu@Sep	10	3.0	120	5.0	0.03 M	~40	[34]
S-doped ZnO	10	0.25	120	6.5	5 mL·L^{-1}	23	[35]
Fe$_3$O$_4$	10	0.25	120	6.5	5 mL·L^{-1}	~60	[35]
Fe$_3$O$_4$@S-doped ZnO	10	0.25	120	6.5	5 mL·L^{-1}	~100	[35]
AC-Fe$_3$O$_4$	12	0.5	60	3.3	20.0 mM	~75	[36]
Fe$_3$O$_4$-CeO$_2$	12	0.5	60	3.3	20.0 mM	~80	[36]
Fe$_3$O$_4$-CeO$_2$/AC	12	0.5	60	3.3	20.0 mM	~98	[36]
Fe@Mpsi	30	1.0	120	Initial pH	2000 mg·L^{-1}	18	[37]
Cu@Mpsi	30	1.0	120	Initial pH	2000 mg·L^{-1}	51	[37]
Fe-Cu@SBA-15	30	1.0	120	Initial pH	2000 mg·L^{-1}	~70	[37]
Fe-Cu@Mpsi	30	1.0	120	Initial pH	2000 mg·L^{-1}	~82	[37]
Fe$_3$O$_4$-zeolite	20	1.0	120	9	5 mL·L^{-1}	88	Present work

2.6.2. Effect of Fe₃O₄ Loading on Zeolite

The amount of Fe_3O_4 loading on zeolite is expected to play a vital role in the overall Fenton oxidation process, since it controls the iron ions concentration and the HO• production rate [38]. For these purposes, the effect of Fe_3O_4 loading on zeolite was studied by varying the amount of Fe_3O_4 from 1.5 to 10 mmol, and the results are shown in Figure 7a. Subsequently, the experimental data were examined using the zero-order, pseudo-first-order, and pseudo-second-order kinetic models and the results are shown in Figure S2. Based on the kinetic data presented in Table S1 and the values of the correlation coefficient, R^2, the Fenton degradation of OFL using FeZ composites follows the pseudo-first-order kinetic model.

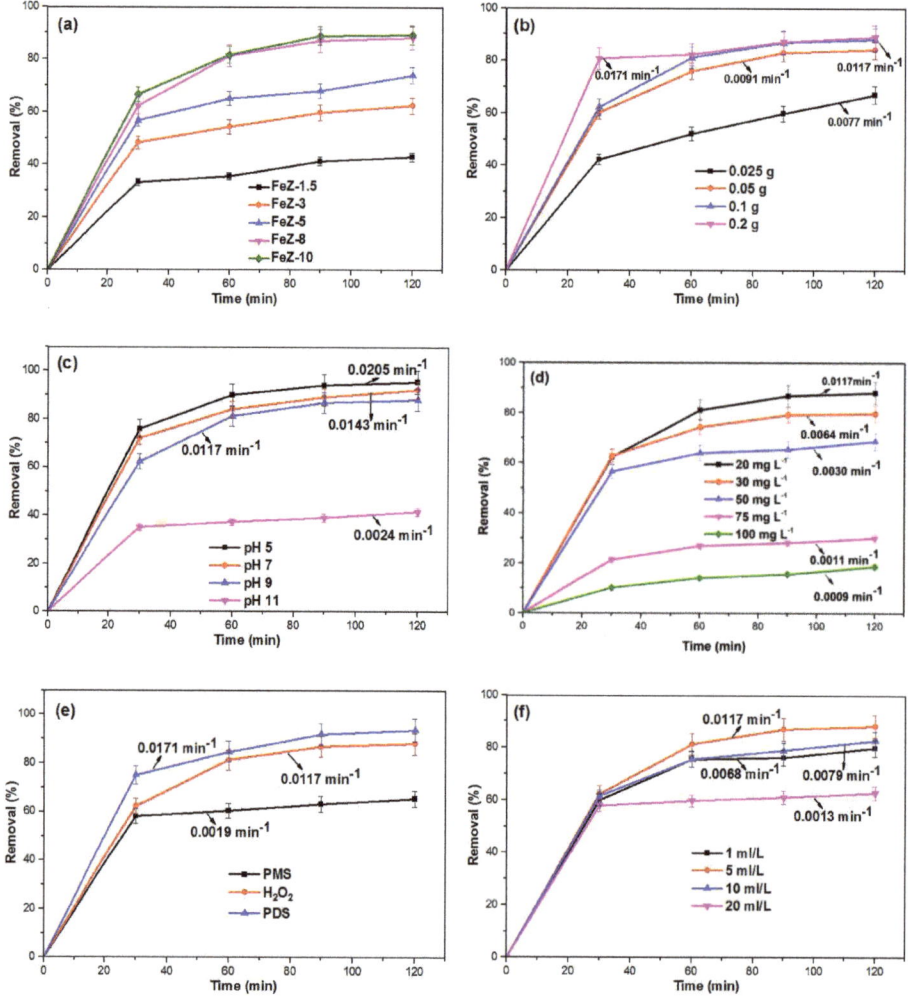

Figure 7. *Cont.*

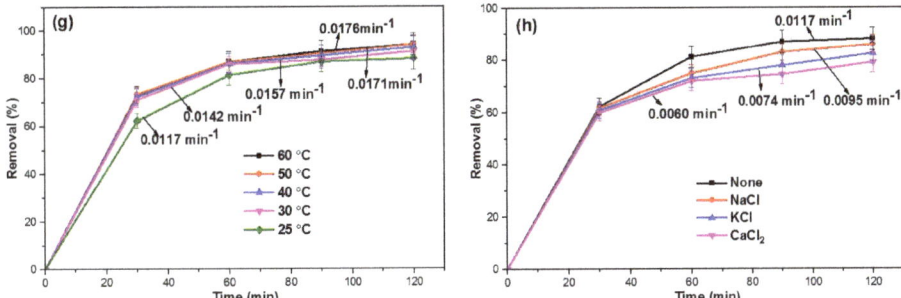

Figure 7. Effects of (**a**) Fe$_3$O$_4$ loading on zeolite, (**b**) FeZ-8 catalyst dosage, (**c**) initial solution pH, (**d**) initial OFL concentration, (**e**) different oxidants, (**f**) H$_2$O$_2$ dosage, (**g**) reaction temperature, and (**h**) inorganic ions on the Fenton degradation of OFL. Unless stated otherwise, reaction conditions are based on: [OFL] = 20 mg·L^{-1}, pH = 9, (FeZ-8) = 1 g·L^{-1}, (H$_2$O$_2$) = 5 mL·L^{-1}.

Furthermore, as the amount of Fe$_3$O$_4$ in the FeZ composite increases from 1.5 to 8 mmol, the degradation rate also increases. Such an effect is attributed to the increase in the production of HO$^{\bullet}$ by higher amount of iron ions. This, however, is contrary to the earlier surface area results presented in Table 2, where FeZ-3 and FeZ-5 (with low loading of Fe$_3$O$_4$) have higher surface areas than FeZ-8 and FeZ-10. This could be explained by the fact that FeZ-8 and FeZ-10 have a higher concentration of iron ions. However, the degradation rate by FeZ-10 is virtually similar to the performance recorded using FeZ-8. Previously, Hassan and Hameed [31] have also reported that at a higher concentration, the excess Fe$_3$O$_4$ would instead scavenge $^{\bullet}$OH. Therefore, for the rest of the study, FeZ-8 was used to investigate the effect of experimental parameters on the degradation of OFL by the FeZ composite. In general, the results from this study have further confirmed the indispensable role of the Fe$_3$O$_4$-zeolite composite catalyst in the Fenton degradation process.

2.6.3. Effect of Catalyst Dosage

The effect of FeZ-8 catalyst dosage was studied using 0.025, 0.05, 0.1, and 0.2 g catalyst loading, and the results are shown in Figure 7b. As the catalyst dosage increases, the degradation efficiency and the rate also increase. Such effect is due to the increased contact of OFL and H$_2$O$_2$ with the high amount of FeZ-8 catalyst [39]. Previously, Shukla, et al. [40] have attributed such performance to the dual effect of combined adsorption and oxidation. However, the increase in degradation efficiency in the same reaction time, when the catalyst dosage was increased from 0.1 to 0.2 g is not appreciable. According to Nguyen, et al. [41], this could be due the fact that at high iron content, the catalyst behaves as an HO$^{\bullet}$ scavenger, and therefore, no significant increase in degradation efficiency could be observed.

2.6.4. Effect of Initial Solution pH

Apart from the performance recorded at the initial solution pH (pH 9.0), the effect of initial solution pH on the Fenton degradation of OFL in FeZ-8 + H$_2$O$_2$ system was determined at pH range of 5–11 and the results are shown in Figure 7c. It is obvious that the percentage of OFL removal efficiency and degradation rate keeps on decreasing with the increase in pH from 5 to 11. This could be explained using the favorable interaction at lower pH values compared to the interaction between the OFL molecules and the FeZ-8 catalyst at higher pH values, since the FeZ-8 catalyst has the pH at point of zero charge (pH$_{zpc}$) of 4.3, and the OFL has the acid dissociation constants values, pKa_1 = 5.98 and pKa_2 = 8.00 [42]. Furthermore, the enhanced generation of HO$^{\bullet}$ at lower pH values could be another contributing factor to the high performance recorded [43]. Meanwhile, the formation of oxyhydroxides like Fe(OH)$_3$ and FeOH^{+} at higher pH values, which have been adsorbed onto the surface of the FeZ-8 catalyst, contributes to the decrease in degradation efficiency with the increase in pH [44]. Yet still, the performance recorded in pH 5–9 is appreciable. The significant decrease in performance at pH 11 is attributed to the decrease in the oxidation potential of HO• and the formation of inactive ferryl ions (FeO^{2+}) at higher pH [45]. However, the result obtained is a tremendous achievement since it shows that the FeZ-8 catalyst could function well within a wide pH range. This is important because the pH range of wastewater is between 5.5 and 9.0 [46].

2.6.5. Effect of Initial OFL Concentration

The effect of initial OFL concentration on the performance of FeZ-8 was studied using OFL solutions with a concentration between 20 and 100 mg·L^{-1}, while other parameters are kept constant, and the results are shown in Figure 7d. The heterogeneous Fenton degradation of OFL using the FeZ-8 catalyst occurs at the surface of the catalyst. It involves the reaction between the generated HO• at the active sites, together with the adsorbed OFL molecules [31]. In this study, since the dose of FeZ-8 and the concentration of H$_2$O$_2$ are constant, the number of hydroxyl radicals produced in the systems is also the same. Thus, the decrease in the degradation efficiency with the increase in OFL concentration is attributed to the low amount of hydroxyl radicals in the system compared to the available OFL molecules.

2.6.6. Effect of Different Oxidants

The effect of different oxidants—peroxydisulfate (PDS), hydrogen peroxide (H$_2$O$_2$), and peroxomonosulfate (PMS)—on the catalytic degradation of OFL was also studied. This study was conducted by activating PDS, PMS, and H$_2$O$_2$ with FeZ-8 to produce $SO_4^{•-}$ and HO•. Based on the results shown in Figure 7e, the order of reactivity is PDS > H$_2$O$_2$ > PMS. The reaction of Fe^{2+} with PS, H$_2$O$_2$, and PMS could be presented in Equations (1)–(3) [47].

$$Fe^{2+} + H_2O_2 \rightarrow Fe^{3+} + OH^- + HO^{•} \tag{1}$$

$$Fe^{2+} + S_2O_8^{2-} \rightarrow Fe^{3+} + SO_4^{•-} + SO_4^{2-} \tag{2}$$

$$Fe^{2+} + HSO_5^- \rightarrow Fe^{3+} + SO_4^{•-} + OH^- \tag{3}$$

In the case of PDS and H$_2$O$_2$, the high performance in FeZ-8/PDS than FeZ-8/H$_2$O$_2$ is due to the rapid formation of reactive oxygen species (ROSs) from PDS activation [48]. In the case of PMS, the inherent pH of aqueous OFL solution is 9. The low performance recorded at this pH could be attributed to the self-dissociation of PMS, which is reported to occur at high pH through non-radical pathways [49].

2.6.7. Effect of H$_2$O$_2$ Dosage

The effect of H$_2$O$_2$ dosage on the Fenton degradation of OFL was studied by varying the amount of H$_2$O$_2$ from 1 to 20 mL·L^{-1} and the results are shown in Figure 7f. From the results obtained, an increase in H$_2$O$_2$ concentration from 1 to 5 mL·L^{-1} resulted in an increase in the Fenton degradation rate from 0.0068 to 0.0118 min^{-1}. Such an effect is attributed to the rise in the number of HO•. However, a further increase in H$_2$O$_2$ concentration from 5 to 20 mL·L^{-1} causes a decrease in Fenton degradation rate to 0.0013 min^{-1}. This could be explained because of the excess H$_2$O$_2$ acting as a hydroxyl radical scavenger to form the less reactive hydroperoxyl radicals (HOO•), which do not contribute much to the oxidation reaction [50].

2.6.8. Effect of Reaction Temperature

The effect of reaction temperature on the Fenton degradation of OFL using the FeZ-8 catalyst has been studied, and the results are depicted in Figure 7g. From the figure, a higher reaction temperature resulted in a little bit higher degradation of OFL, which is in agreement with the report by Soon and Hameed [39]. The authors stated that higher temperature accelerates the decomposition of H$_2$O$_2$

into HO•. In general, the findings show that the Fenton degradation of OFL by the FeZ-8 catalyst was not very sensitive to the reaction temperature. Furthermore, the activation energy (E_a) for the Fenton degradation of OFL using the FeZ-8 catalyst was estimated using the plot of ln k against 1/T, based on the Arrhenius equation ln k = ln A − E_a/RT, where k is the rate constant, R is the universal gas constant (8.314 J mol^{-1} K^{-1}), and A is pre-exponential factor, and the results are shown in Figure S3. Using the slope of the plot, the E_a value was obtained as 8.8860 kJ mol^{-1} and thus, indicates low reaction activation energy [51]. Such a value, which is significantly lower than the 213.8 kJ mol^{-1} for the dissociation of H_2O_2 to HO• [52], confirms the catalytic property of FeZ-8.

2.6.9. Effect of Inorganic Salts

Apart from organic pollutants, real wastewater often contains some inorganic salts, which could possibly affect the efficiency of the degradation process during the Fenton reaction. For this purpose, 10 mmol of NaCl, KCl, and $CaCl_2$ salts were separately introduced during the Fenton degradation of OFL, and the results are shown in Figure 7h. Although no significant decrease in degradation efficiency of OFL was noticed upon the introduction of the various inorganic salts, the effect was more pronounced in the presence of $CaCl_2$. Such a minor decrease in degradation efficiency could be attributed to the possible adsorption of Cl$^-$ on the surface of FeZ-8 catalyst, thereby blocking some active sites [53] or because Cl$^-$ might act as a hydroxyl radicals scavenger, as shown in Equation (4) below [54,55]:

$$Cl^- + {}^\bullet OH \rightarrow Cl^\bullet + OH^- \qquad (4)$$

2.6.10. TOC Removal Studies

The catalytic activity of the FeZ-8 catalyst was further investigated by determining the total organic carbon (TOC) concentration of the OFL solution after the establishment of the adsorption–desorption equilibrium and Fenton degradation for 120 min, and the results are shown in Figure 8. Although the TOC removal efficiency was approximately 40% after the establishment of the adsorption–desorption equilibrium, the efficiency reached more than 50% after 120 min of Fenton degradation reaction. Such results confirmed that the FeZ-8 catalyst is capable of mineralizing the fluoroquinolone antibiotic.

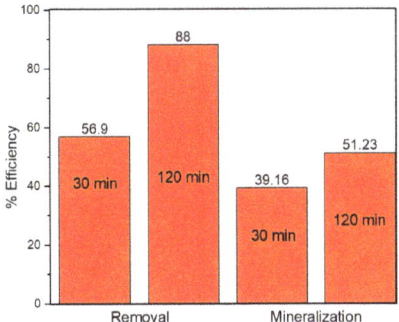

Figure 8. Total organic carbon (TOC) removal using FeZ-8.

2.6.11. Hetero-Fenton Degradation Mechanism

The reaction pathway for OFL degradation using the FeZ composite as a hetero-Fenton catalyst has been proposed and is schematically shown in Figure 9. The first stage involves the in-dark adsorption of OFL onto the surface of the FeZ composite catalyst. The Fe_3O_4 in FeZ is composed of Fe^{2+} and Fe^{3+}. The Fe^{2+} ions react with H_2O_2 to form HO• and more Fe^{3+} (Equation (5)). The HO• is responsible for the degradation of OFL into CO_2, H_2O, and other intermediates. However, the H_2O_2 further reduces the Fe^{3+} back to Fe^{2+} (Equation (7)) and the cycle continues.

$$Fe^{2+} + H_2O_2 \rightarrow Fe^{3+} + OH^- + HO^\bullet \qquad (5)$$

$$Fe^{3+} + H_2O_2 \rightarrow Fe^{2+} + H^+ + HO_2^\bullet \qquad (6)$$

$$OFL + OH^\bullet \rightarrow CO_2 + H_2O + intermediates \qquad (7)$$

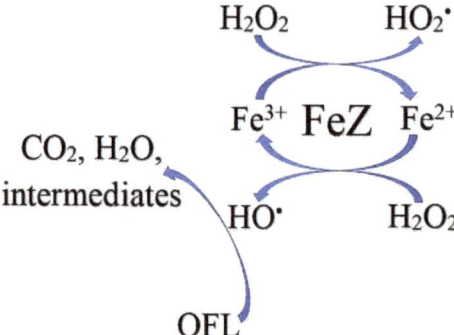

Figure 9. Proposed reaction pathway for OFL degradation using FeZ composite as hetero-Fenton catalyst.

2.6.12. Reusability and Stability Studies

A reusability study was conducted to investigate the stability of the FeZ-8 catalyst during repeated cycles. As shown in Figure 10a, the performance of the FeZ-8 catalyst after the fifth cycle remained significant, as the marginal drop in the efficiency is still below 10%, an indication that the FeZ-8 catalyst is reusable. The stability of the FeZ-8 catalyst was confirmed by examining the FTIR spectra of the catalyst before and after five cycles of repeated Fenton degradation studies, and no obvious difference exists between the two FTIR spectra of the FeZ-8 catalyst.

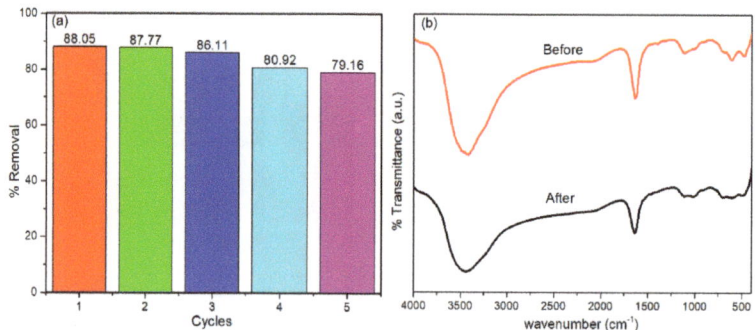

Figure 10. (**a**) Reusability studies of Fenton degradation of OFL using FeZ-8. (**b**) FTIR spectra of FeZ-8 catalyst before and after the Fenton degradation of OFL.

3. Materials and Methods

3.1. Chemicals

Ferric chloride hexahydrate ($FeCl_3 \cdot 6H_2O$), sodium chloride (NaCl), and calcium chloride ($CaCl_2$) were obtained from Bendosen (Shah Alam, Selangor, Malaysia). Hydrogen peroxide (H_2O_2), nitric acid (HNO_3), ethanol (CH_3CH_2OH), and sodium hydroxide (NaOH) were obtained from QReC Chemicals (Rawang; Selangor Malaysia). Sodium peroxydisulfate (PDS, $Na_2S_2O_8$), hydrochloric acid (HCl), and ammonia solution were obtained from Merck Chemicals (Darmstadt, Germany). Iron (II) sulfate heptahydrate ($FeSO_4 \cdot 7H_2O$) and zeolite were obtained from Sigma-Aldrich (Saint Louis, MO, USA). Potassium chloride (KCl) was obtained from Fluka (St. Gallen, Switzerland), while potassium peroxomonosulfate (PMS, $2KHSO_5 \cdot KHSO_4 \cdot K_2SO_4$) was supplied by Acros Organics (Morris, NJ, USA). All chemicals were of analytical grade and deionized water was used throughout.

3.2. Preparation of FeZ Composites

Fe_3O_4-zeolite (FeZ) composites were synthesized using the co-precipitation method [23]. Briefly, equimolar amounts (1.5, 3, 5, 8, or 10 mmol) of $FeSO_4 \cdot 7H_2O$ and $FeCl3 \cdot 6H2O$ were dissolved in 60 mL of 10 mmol L^{-1} aqueous HCl solution and heated to 80 °C (Solution A). Subsequently, 0.5 g of zeolite was introduced into solution A and stirred for 1 h (Solution B). Then, 40 mL of ammonia solution was added slowly into solution B and stirred together for 2 h, to form FeZ composites. Finally, the product was filtered, washed several times with distilled water and absolute ethanol, before being dried in an oven at 60 °C overnight. The composites were named FeZ-1.5, FeZ-3, FeZ-5, FeZ-8, and FeZ-10, based on the equimolar amount of iron precursors used during the synthesis.

3.3. Characterizations

The morphologies of the catalysts were determined using a transmission electron microscope (TEM) (model: Technai G2 F20, Eindhoven, Netherlands) and field emission scanning electron microscope (model: Leo Supra 50 VP FESEM, Eindhoven, Netherlands) accompanied with an energy dispersive X-ray detector (EDX) to determine the elemental composition of the catalysts. FTIR spectra were recorded with KBr pressed disks using a Perkin Elmer 2000 spectrometer (Beaconsfied, England). X-ray photoelectron spectroscopy (XPS) analysis was carried out using an ULVAC-PHI Quantera II XPS system with a monochromatic Al K-Alpha source (Chigasaki, Japan). The specific surface area of the catalysts was measured by the Brunauer–Emmett–Teller (BET) method using a Micromeritics ASAP (model 2020 V 4.01) surface area and porosity analyzer (Norcross, GA, USA).

3.4. Heterogeneous Fenton Degradation of OFL

The catalytic activity of FeZ composites was evaluated by degrading OFL in the presence of H_2O_2 in ambient conditions. In a typical procedure, 0.1 g of the catalyst was added into a 100 mL (20 ppm) aqueous OFL solution. The mixture was continuously stirred for 30 min in the dark in order to disperse FeZ catalyst and establish an adsorption–desorption equilibrium. Subsequently, 0.5 mL of aqueous H_2O_2 solution (30% w/w) was added to the mixture so as to enhance the catalytic performance via Fenton reaction. At a regular time interval, 5 mL of the reaction suspension was sampled and centrifuged immediately to separate the catalyst from the solution. The remnant OFL concentration in the supernatant was then determined at 286 nm using a Shimadzu UV 2600 spectrophotometer (version 1.03) operating using a UV probe 2.42. The effects of experimental conditions such as Fe_3O_4 loading on zeolite, catalyst loading, initial solution pH, initial OFL concentration, different oxidants, H_2O_2 dosage, reaction temperature, and inorganic salts were studied in order to determine the performance of FeZ catalyst towards the degradation of OFL under different conditions. The amount of Fe_3O_4 loaded on zeolite was varied from 1.5 to 10 mmol; catalyst dosage was varied from 0.25 to 2 g·L^{-1}; initial solution pH was adjusted between 5 and 11 using HNO_3 or NaOH; initial OFL concentration was varied between 20 and 100 ppm; different oxidants such as potassium peroxomonosulfate, sodium peroxydisulfate, and hydrogen peroxide; dosage of H_2O_2 from 1 to 20 mL·L^{-1}; reaction temperature between 25 and 60 °C; and different inorganic salts such as NaCl, KCl, and CaCl$_2$. The efficiency of the FeZ composite towards the removal (both adsorption and Fenton degradation) of OFL was calculated using Equation (8):

$$\% \, Removal = \frac{C_o - C_t}{C_o} \times 100\% \tag{8}$$

where C_o and C_t are the initial and the concentration of OFL solution at time t. Total organic carbon (TOC) was determined using a Shimadzu TOC-L analyzer. Finally, the reusability and stability of the catalyst were also studied.

4. Conclusions

Fe$_3$O$_4$-zeolite (FeZ) composites were synthesized via a facile co-precipitation method and were subsequently characterized using XPS, FTIR, FESEM-EDX, TEM, and BET, which successfully confirmed the immobilization of Fe$_3$O$_4$ onto zeolite. The immobilization of Fe$_3$O$_4$ onto zeolite resulted in the formation of FeZ composites with higher surface area and pore volume, compared to pristine zeolite. The effects of different operational parameters such as Fe$_3$O$_4$ loading on zeolite, catalyst loading, initial solution pH, initial OFL concentration, different oxidants, H$_2$O$_2$ dosage, reaction temperature, and inorganic salts on OFL degradation were systematically investigated. Among all the FeZ composites, the FeZ-8 composite exhibited a good catalytic capacity for the hetero-Fenton degradation of OFL. Under the reaction conditions of 1 g·L^{-1} catalyst dosage, pH of 9, and H$_2$O$_2$ dosage of 5mL·L^{-1}, 88% of OFL degradation efficiency and 51.3% of TOC removal efficiency were achieved using the FeZ-8 composite. Likewise, results from reusability tests revealed only a slight decrease in degradation ability after five runs. Therefore, the FeZ-8 composite could be a promising hetero-Fenton catalyst for the treatment of wastewater contaminated with antibiotics. A suitable hetero-Fenton degradation pathway has been proposed.

Supplementary Materials: The following are available online at http://www.mdpi.com/2073-4344/10/11/1241/s1, Figure S1: N2 adsorption-desorption isotherms and the corresponding pore size distribution (insets) for zeolite (a), FeZ-1.5 (b), FeZ-3 (c), FeZ-5 (d), FeZ-8 (e) and FeZ-10 (f) composites, Figure S2: (a) (C$_0$–C) versus t plots based on the zero-order kinetics model; (b) ln(C$_0$/C) versus t plots based on first-order kinetics model and (c) 1/C–1/C$_0$ versus t plots based on second order kinetics model for the effect of Fe$_2$O$_3$ loading on zeolite on the degradation of OFL, Table S1: Parameters of linear regression for different kinetic models in FeZ/H$_2$O$_2$ systems,

Author Contributions: Conceptualization, methodology, resources, and writing—review and editing, R.A.; writing—review and editing, W.D.O. and S.S.I.; investigation and writing—original draft preparation, A.R.D.A. All authors have read and agreed to the published version of the manuscript.

Funding: The authors gratefully acknowledged the financial support from Universiti Sains Malaysia under RUI Grant No. 1001/PKIMIA/8011117.

Conflicts of Interest: The authors declare no conflict of interest. The funders had no role in the design of the study; in the collection, analyses, or interpretation of data; in the writing of the manuscript, or in the decision to publish the results.

References

1. Wang, W.; Zhai, C.; Peng, Y.; Chao, K. A Nondestructive Detection Method for Mixed Veterinary Drugs in Pork Using Line-Scan Raman Chemical Imaging Technology. *Food Anal. Methods* **2019**, *12*, 658–667. [CrossRef]
2. Sharma, S.; Bhandari, A.; Choudhary, V.; Rajpurohit, H.; Khandelwal, P. RP-HPLC method for simultaneous estimation of nitazoxanide and ofloxacin in tablets. *Ind. J. Pharm. Sci.* **2011**, *73*, 84. [CrossRef]
3. Janos, F.; Robin, G.C. *Analogue-based Drug Discovery*, 1st ed.; Wiley-VCH: Hoboken, NJ, USA, 2006; ISBN 9783527607495.
4. Sun, J.; Song, M.; Feng, J.; Pi, Y. Highly efficient degradation of ofloxacin by UV/Oxone/Co^{2+} oxidation process. *Environ. Sci. Pollut. Res.* **2012**, *19*, 1536–1543. [CrossRef]

5. Chen, T.S.; Kuo, Y.M.; Chen, J.L.; Huang, K.L. Anodic degradation of ofloxacin on a boron-doped diamond electrode. *Int. J. Electrochem. Sci.* **2013**, *8*, 7625–7633.
6. Elfiky, M.; Salahuddin, N.; Hassanein, A.; Matsuda, A.; Hattori, T. Detection of antibiotic Ofloxacin drug in urine using electrochemical sensor based on synergistic effect of different morphological carbon materials. *Microchem. J.* **2019**, *146*, 170–177. [CrossRef]
7. Peres, M.; Maniero, M.; Guimarães, J. Photocatalytic degradation of ofloxacin and evaluation of the residual antimicrobial activity. *Photochem. Photobiol. Sci.* **2015**, *14*, 556–562. [CrossRef]
8. Enick, O.V.; Moore, M.M. Assessing the assessments: Pharmaceuticals in the environment. *Environ. Impact Assess. Rev.* **2007**, *27*, 707–729. [CrossRef]
9. Esposito, B.R.; Capobianco, M.L.; Martelli, A.; Navacchia, M.L.; Pretali, L.; Saracino, M.; Zanelli, A.; Emmi, S.S. Advanced water remediation from ofloxacin by ionizing radiation. *Radiat. Phys. Chem.* **2017**, *141*, 118–124. [CrossRef]
10. Wuana, R.A.; Sha'Ato, R.; Iorhen, S. Aqueous phase removal of ofloxacin using adsorbents from Moringa oleifera pod husks. *Adv. Environ. Res.* **2015**, *4*, 49–68. [CrossRef]
11. La Farre, M.; Pérez, S.; Kantiani, L.; Barceló, D. Fate and toxicity of emerging pollutants, their metabolites and transformation products in the aquatic environment. *TrAC Trends Anal. Chem.* **2008**, *27*, 991–1007. [CrossRef]
12. Dong, R.; Yu, G.; Guan, Y.; Wang, B.; Huang, J.; Deng, S.; Wang, Y. Occurrence and discharge of pharmaceuticals and personal care products in dewatered sludge from WWTPs in Beijing and Shenzhen. *Emerg. Contam.* **2016**, *2*, 1–6. [CrossRef]
13. Zhang, T. Antibiotics and resistance genes in wastewater treatment plants. *AMR Control* **2016**, *9*, 11–20.
14. Varjani, S.J.; Sudha, M.C. Treatment Technologies for Emerging Organic Contaminants Removal from Wastewater. In *Water Remediation*; Springer Nature: Singapore, 2018; pp. 91–115.
15. Cuerda-Correa, E.M.; Alexandre-Franco, M.F.; Fernández-González, C. Advanced oxidation processes for the removal of antibiotics from water. An overview. *Water* **2020**, *12*, 102. [CrossRef]
16. Jaafarzadeh, N.; Kakavandi, B.; Takdastan, A.; Kalantary, R.R.; Azizi, M.; Jorfi, S. Powder activated carbon/Fe_3O_4 hybrid composite as a highly efficient heterogeneous catalyst for Fenton oxidation of tetracycline: Degradation mechanism and kinetic. *RSC Adv.* **2015**, *5*, 84718–84728. [CrossRef]
17. Cleveland, V.; Bingham, J.P.; Kan, E. Heterogeneous Fenton degradation of bisphenol A by carbon nanotube-supported Fe_3O_4. *Sep. Purif. Technol.* **2014**, *133*, 388–395. [CrossRef]
18. Hua, Z.; Ma, W.; Bai, X.; Feng, R.; Yu, L.; Zhang, X.; Dai, Z. Heterogeneous Fenton degradation of bisphenol A catalyzed by efficient adsorptive Fe_3O_4/GO nanocomposites. *Environ. Sci. Pollut. Res.* **2014**, *21*, 7737–7745. [CrossRef]
19. Mazilu, I.; Ciotonea, C.; Chirieac, A.; Dragoi, B.; Catrinescu, C.; Ungureanu, A.; Petit, S.; Royer, S.; Dumitriu, E. Synthesis of highly dispersed iron species within mesoporous (Al-) SBA-15 silica as efficient heterogeneous Fenton-type catalysts. *Microporous Mesoporous Mater.* **2017**, *241*, 326–337. [CrossRef]
20. Krishna, L.S.; Soontarapa, K.; Asmel, N.K.; Kabir, M.A.; Yuzir, A.; Yaacob, W.Z.W.; Sarala, Y. Adsorption of acid blue 25 from aqueous solution using zeolite and surfactant modified zeolite. *Desalin. Water Treat.* **2019**, *150*, 348–360. [CrossRef]
21. De Sousa, D.N.R.; Insa, S.; Mozeto, A.A.; Petrovic, M.; Chaves, T.F.; Fadini, P.S. Equilibrium and kinetic studies of the adsorption of antibiotics from aqueous solutions onto powdered zeolites. *Chemosphere* **2018**, *205*, 137–146. [CrossRef]
22. Zhu, D.; Liu, S.; Chen, M.; Zhang, J.; Wang, X. Flower-like-flake Fe_3O_4/g-C_3N_4 nanocomposite: Facile synthesis, characterization, and enhanced photocatalytic performance. *Colloids Surf. A Physicochem. Eng. Asp.* **2018**, *537*, 372–382. [CrossRef]
23. Huang, R.; Fang, Z.; Yan, X.; Cheng, W. Heterogeneous sono-Fenton catalytic degradation of bisphenol A by Fe_3O_4 magnetic nanoparticles under neutral condition. *Chem. Eng. J.* **2012**, *197*, 242–249. [CrossRef]
24. Adimoolam, M.G.; Amreddy, N.; Nalam, M.R.; Sunkara, M.V. A simple approach to design chitosan functionalized Fe_3O_4 nanoparticles for pH responsive delivery of doxorubicin for cancer therapy. *J. Magn. Magn. Mater.* **2018**, *448*, 199–207. [CrossRef]
25. Caglar, B.; Guner, E.K.; Keles, K.; Özdokur, K.V.; Cubuk, O.; Coldur, F.; Caglar, S.; Topcu, C.; Tabak, A. Fe_3O_4 nanoparticles decorated smectite nanocomposite: Characterization, photocatalytic and electrocatalytic activities. *Solid State Sci.* **2018**, *83*, 122–136. [CrossRef]

26. Zhao, H.; Weng, L.; Cui, W.W.; Zhang, X.R.; Xu, H.Y.; Liu, L.Z. In situ anchor of magnetic Fe_3O_4 nanoparticles onto natural maifanite as efficient heterogeneous Fenton-like catalyst. *Front. Mat. Sci.* **2016**, *10*, 300–309. [CrossRef]
27. Karami, B.; Hoseini, S.J.; Eskandari, K.; Ghasemi, A.; Nasrabadi, H. Synthesis of xanthene derivatives by employing Fe_3O_4 nanoparticles as an effective and magnetically recoverable catalyst in water. *Catal. Sci. Technol.* **2012**, *2*, 331–338. [CrossRef]
28. Zhao, T.; Ji, X.; Guo, X.; Jin, W.; Dang, A.; Li, H.; Li, T. Preparation and electrochemical property of Fe_3O_4/MWCNT nanocomposite. *Chem. Phy. Lett.* **2016**, *653*, 202–206. [CrossRef]
29. Liu, Y.; Xu, J.; Wang, L.; Zhang, H.; Xu, P.; Duan, X.; Sun, H.; Wang, S. Three-dimensional BiOI/BiOX (X = Cl or Br) nanohybrids for enhanced visible-light photocatalytic activity. *Nanomaterials* **2017**, *7*, 64. [CrossRef]
30. Liao, Q.; Sun, J.; Gao, L. Degradation of phenol by heterogeneous Fenton reaction using multi-walled carbon nanotube supported Fe_2O_3 catalysts. *Colloids Surf. A Physicochem. Eng. Asp.* **2009**, *345*, 95–100. [CrossRef]
31. Hassan, H.; Hameed, B. Fe–clay as effective heterogeneous Fenton catalyst for the decolorization of Reactive Blue 4. *Chem. Eng. J.* **2011**, *171*, 912–918. [CrossRef]
32. Hu, X.; Liu, B.; Deng, Y.; Chen, H.; Luo, S.; Sun, C.; Yang, P.; Yang, S. Adsorption and heterogeneous Fenton degradation of 17α-methyltestosterone on nano Fe_3O_4/MWCNTs in aqueous solution. *Appl. Catal. B Environ.* **2011**, *107*, 274–283. [CrossRef]
33. Lan, H.; Wang, A.; Liu, R.; Liu, H.; Qu, J. Heterogeneous photo-Fenton degradation of acid red B over Fe_2O_3 supported on activated carbon fiber. *J. Hazard. Mat.* **2015**, *285*, 167–172. [CrossRef]
34. Tian, Y.; He, X.; Zhou, H.; Tian, X.; Nie, Y.; Zhou, Z.; Yang, C.; Li, Y. Efficient fenton-like degradation of ofloxacin over bimetallic Fe–Cu@ Sepiolite composite. *Chemosphere* **2020**, *257*, 127209. [CrossRef]
35. Wang, X.; Jin, H.; Wu, D.; Nie, Y.; Tian, X.; Yang, C.; Zhou, Z.; Li, Y. Fe_3O_4@S-doped ZnO: A magnetic, recoverable, and reusable Fenton-like catalyst for efficient degradation of ofloxacin under alkaline conditions. *Environ. Res.* **2020**, *186*, 109626. [CrossRef]
36. Liu, J.; Wu, X.; Liu, J.; Zhang, C.; Hu, Q.; Hou, X. Ofloxacin degradation by Fe_3O_4-CeO_2/AC Fenton-like system: Optimization, kinetics, and degradation pathways. *Mole. Cat.* **2019**, *465*, 61–67. [CrossRef]
37. Zheng, C.; Yang, C.; Cheng, X.; Xu, S.; Fan, Z.; Wang, G.; Wang, S.; Guan, X.; Sun, X. Specifically enhancement of heterogeneous Fenton-like degradation activities for ofloxacin with synergetic effects of bimetallic Fe-Cu on ordered mesoporous silicon. *Sep. Purif. Technol.* **2017**, *189*, 357–365. [CrossRef]
38. Azizan, Z.; Azran, M.; Hassan, H.; Faraziehan, S.; Abu Hassan, N. Decolorization of Reactive Black 5 using Fe-areca nut as a heterogeneous Fenton catalyst. *Proc. Appl. Mech. Mater.* **2014**, *661*, 29–33. [CrossRef]
39. Soon, A.N.; Hameed, B. Degradation of Acid Blue 29 in visible light radiation using iron modified mesoporous silica as heterogeneous Photo-Fenton catalyst. *Appl. Catal. A Gen.* **2013**, *450*, 96–105. [CrossRef]
40. Shukla, P.; Wang, S.; Sun, H.; Ang, H.-M.; Tadé, M. Adsorption and heterogeneous advanced oxidation of phenolic contaminants using Fe loaded mesoporous SBA-15 and H_2O_2. *Chem. Eng. J.* **2010**, *164*, 255–260. [CrossRef]
41. Nguyen, T.D.; Phan, N.H.; Do, M.H.; Ngo, K.T. Magnetic Fe_2MO_4 (M: Fe, Mn) activated carbons: Fabrication, characterization and heterogeneous Fenton oxidation of methyl orange. *J. Hazard. Mat.* **2011**, *185*, 653–661. [CrossRef] [PubMed]
42. Yu, L.; Chen, J.; Liang, Z.; Xu, W.; Chen, L.; Ye, D. Degradation of phenol using Fe_3O_4-GO nanocomposite as a heterogeneous photo-Fenton catalyst. *Sep. Purif. Technol.* **2016**, *171*, 80–87. [CrossRef]
43. Li, W.; Wan, D.; Wang, G.; Chen, K.; Hu, Q.; Lu, L. Heterogeneous Fenton degradation of Orange II by immobilization of Fe_3O_4 nanoparticles onto Al-Fe pillared bentonite. *Korean J. Chem. Eng.* **2016**, *33*, 1557–1564. [CrossRef]
44. Tang, X.; Feng, Q.; Liu, K.; Li, Z.; Wang, H. Fabrication of magnetic Fe_3O_4/silica nanofiber composites with enhanced Fenton-like catalytic performance for Rhodamine B degradation. *J. Mat. Sci.* **2018**, *53*, 369–384. [CrossRef]
45. Shi, X.; Tian, A.; You, J.; Yang, H.; Wang, Y.; Xue, X. Degradation of organic dyes by a new heterogeneous Fenton reagent-Fe_2GeS_4 nanoparticle. *J. Hazard. Mat.* **2018**, *353*, 182–189. [CrossRef] [PubMed]
46. Wang, Y.; Lu, J.; Wu, J.; Liu, Q.; Zhang, H.; Jin, S. Adsorptive removal of fluoroquinolone antibiotics using bamboo biochar. *Sustainability* **2015**, *7*, 12947–12957. [CrossRef]

47. Khan, J.A.; He, X.; Khan, H.M.; Shah, N.S.; Dionysiou, D.D. Oxidative degradation of atrazine in aqueous solution by UV/H_2O_2/Fe^{2+}, UV/$S_2O_8^{2-}$/Fe^{2+} and UV/HSO_5^-/Fe^{2+} processes: A comparative study. *Chem. Eng. J.* **2013**, *218*, 376–383. [CrossRef]
48. Du, J.; Wang, Y.; Xu, T.; Zheng, H.; Bao, J. Synergistic degradation of PNP via coupling H_2O_2 with persulfate catalyzed by nano zero valent iron. *RSC Adv.* **2019**, *9*, 20323–20331. [CrossRef]
49. Ji, Y.; Wang, L.; Jiang, M.; Yang, Y.; Yang, P.; Lu, J.; Ferronato, C.; Chovelon, J.M. Ferrous-activated peroxymonosulfate oxidation of antimicrobial agent sulfaquinoxaline and structurally related compounds in aqueous solution: Kinetics, products, and transformation pathways. *Environ. Sci. Pollut. Res.* **2017**, *24*, 19535–19545. [CrossRef]
50. Liu, Y.; Jin, W.; Zhao, Y.; Zhang, G.; Zhang, W. Enhanced catalytic degradation of methylene blue by α-Fe_2O_3/graphene oxide via heterogeneous photo-Fenton reactions. *Appl. Catal. B Environ.* **2017**, *206*, 642–652. [CrossRef]
51. Feng, J.; Hu, X.; Yue, P.L. Novel bentonite clay-based Fe- nanocomposite as a heterogeneous catalyst for photo-fenton discoloration and mineralization of orange II. *Environ. Sci. Technol.* **2004**, *38*, 269–275. [CrossRef]
52. Chen, F.; Xie, S.; Huang, X.; Qiu, X. Ionothermal synthesis of Fe_3O_4 magnetic nanoparticles as efficient heterogeneous Fenton-like catalysts for degradation of organic pollutants with H_2O_2. *J. Hazard. Mat.* **2017**, *322*, 152–162. [CrossRef]
53. Wen, X.J.; Niu, C.G.; Zhang, L.; Liang, C.; Zeng, G.M. An in depth mechanism insight of the degradation of multiple refractory pollutants via a novel $SrTiO_3$/BiOI heterojunction photocatalysts. *J. Catal.* **2017**, *356*, 283–299. [CrossRef]
54. Tsuneda, S.; Ishihara, Y.; Hamachi, M.; Hirata, A. Inhibition effect of chlorine ion on hydroxyl radical generation in UV-H_2O_2 process. *Water Sci. Technol.* **2002**, *46*, 33–38. [CrossRef]
55. Dugandžić, A.M.; Tomašević, A.V.; Radišić, M.M.; Šekuljica, N.Ž.; Mijin, D.Ž.; Petrović, S.D. Effect of inorganic ions, photosensitisers and scavengers on the photocatalytic degradation of nicosulfuron. *J. Photochem. Photobiol. A Chem.* **2017**, *336*, 146–155. [CrossRef]

Publisher's Note: MDPI stays neutral with regard to jurisdictional claims in published maps and institutional affiliations.

© 2020 by the authors. Licensee MDPI, Basel, Switzerland. This article is an open access article distributed under the terms and conditions of the Creative Commons Attribution (CC BY) license (http://creativecommons.org/licenses/by/4.0/).

Article

Glycine–Nitrate Combustion Synthesis of Cu-Based Nanoparticles for NP9EO Degradation Applications

Hsu-Hui Cheng [1,*], Shiao-Shing Chen [2,*], Hui-Ming Liu [3], Liang-Wei Jang [2] and Shu-Yuan Chang [4]

1. School of Environmental and Chemical Engineering, Zhaoqing University, Zhaoqing 526061, China
2. Institute of Environment Engineering and Management, National Taipei University of Technology, 1, Sec.3 Chung-Hsiao E. Rd., Taipei 10643, Taiwan; nzblues81156@gmail.com
3. Department of Safety, Health and Environmental Engineering, Hungkuang University, Taichung 43302, Taiwan; hmliu@sunrise.hk.edu.tw
4. School of Foreign Languages, Zhaoqing University, Zhaoqing 526061, China; elliechang0334@gmail.com
* Correspondence: hhcheng1126@gmail.com (H.-H.C.); f10919@ntut.edu.tw (S.-S.C.); Tel.: +86-198-6738-2107 (H.-H.C.)

Received: 15 July 2020; Accepted: 1 September 2020; Published: 15 September 2020

Abstract: Copper-based nanoparticles were synthesized using the glycine–nitrate process (GNP) by using copper nitrate trihydrate [$Cu(NO_3)_2 \cdot 3H_2O$] as the main starting material, and glycine [$C_2H_5NO_2$] as the complexing and incendiary agent. The as-prepared powders were characterized through X-ray diffraction (XRD), Brunauer–Emmett–Teller (BET), X-ray photoelectron spectroscopy (XPS), and scanning electron microscopy analysis. Using $Cu(NO_3)_2 \cdot 3H_2O$ as the oxidizer (N) and glycine as fuel (G), we obtained CuO, mixed-valence copper oxides ($CuO + Cu_2O$, G/N = 0.3–0.5), and metallic Cu (G/N = 0.7). The XRD and BET results indicated that increasing the glycine concentration (G/N = 0.7) and reducing the particle surface area increased the yield of metallic Cu. The effects of varying reaction parameters, such as catalyst activity, catalyst dosage, and H_2O_2 concentration on nonylphenol-9-polyethoxylate (NP9EO) degradation, were assessed. With a copper-based catalyst in a heterogeneous system, the NP9EO and total organic carbon removal efficiencies were 83.1% and 70.6%, respectively, under optimum operating conditions (pH, 6.0; catalyst dosage, 0.3 g/L; H_2O_2 concentration, 0.05 mM). The results suggest that the removal efficiency increased with an increase in H_2O_2 concentration but decreased when the H_2O_2 concentration exceeded 0.05 mM. Furthermore, the trend of photocatalytic activity was as follows: G/N = 0.5 > G/N = 0.7 > G/N = 0.3. The G/N = 0.5 catalysts showed the highest photocatalytic activity and resulted in 94.6% NP9EO degradation in 600 min.

Keywords: glycine–nitrate process; copper-based nanoparticles; photocatalysis activity; NP9EO

1. Introduction

In recent years, many studies have attempted to develop nanomaterials through green synthesis to solve environmental issues. To synthesize nanoparticles with a high, specific surface area, it is also very important to control the size distribution of ultrafine particles [1]. The glycine–nitrate process (GNP) is effective, simple, and environmentally compatible for the efficient synthesis of various nanostructured metal oxides [2–5]. The GNP is highly suitable for synthesizing fine crystalline powders and nanometer-scale particles with desirable characteristics, such as high chemical homogeneity, high purity, and a narrow size distribution range of nanoparticles [6]. The GNP is a type of solution combustion synthesis, which involves a self-sustained reaction between an oxidizer, such as a metal nitrate solution, and a fuel, such as glycine [5,7,8]. Furthermore, glycine supplies the energy required for combustion and functions as a complexing agent. An advantage of the GNP is that the rapid evolution of a large volume of gaseous products during combustion dissipates heat from the process and limits temperature increases, which reduces the probability of local sintering among particles while facilitating fine powder formation [6].

Nanosized transition metal oxide (TMO) particles have recently attracted considerable attention because they have unique applications with various types of matter, such as catalysts [2,5], photocatalysts [8], solar cells [5], and gas sensors [9]. Preparation of high-quality nanostructures with defined and controllable size and morphology is the main determinant of catalytic performance [10]. Among nanocatalysts, TMO-based catalysts offer active sites with low activation potential for hydrogen evolution systems [11]; in particular, noble metals (e.g., gold, silver, and platinum) are widely used to optimize catalytic efficiency. However, the high cost of noble metals remains a major obstacle in the development of stable and highly efficient catalysts. Consequently, reducing the amount of noble metals without compromising their catalytic performance is an urgent requirement for catalysis.

Compared with noble metal catalysts, copper is relatively inexpensive and moderately abundant [12]. Copper oxides (Cu_xO_y, where x:y = 1:1 or 1:2) and metallic copper are widely used as TMO catalysts. Copper oxides are heterogeneous catalysts with high recyclability, high catalytic efficiency, and low band-gap energies [10]. Cu oxides are used as TMO catalysts; Cu_xO_y catalysts are p-type semiconductors with a monoclinic structure, a bandgap of 1.7–2.17 eV [13], and a high theoretical photocurrent density of -14.7 mA/cm^2 [14,15]; therefore, they are suitable for application in the photocatalysis field. Although these superior properties of copper oxide have been of great interest to researchers, the transformations of cubic copper (I) oxide (Cu_2O) to copper (II) oxide (CuO) and Cu are still unavoidable due to the thermodynamically feasible parasitic reactions of self-oxidation and self-reduction [16]. Nevertheless, the presence of both Cu^+ and Cu^{2+} in oxides has been shown to be beneficial in some cases [17]. In all applications of Cu_xO_y, morphology, structure, size, and surface area are the main determinants of its catalytic performance [15,18]. Compared with other TMOs, such as titanium dioxide (TiO_2), zinc oxide (ZnO), iron oxide (Fe_2O_3), and cadmium sulfide (CdS) [18], only a few studies have described synthesis strategies for different Cu_xO_y nanoparticles along with the introduction of their related applications.

Nonylphenol ethoxylates (NPEOs) are nonionic surfactants that belong to the alkylphenol ethoxylates (APEOs) family, and are used primarily in emulsifiers, detergents, wetting agents, and dispersing agents, etc. Of these, NPEOs with an average of nine ethylene oxide units (NP9EO) are well known for their commercial application [19]. The biodegradation of NPEOs in the environment leads to the formation of nonylphenol, which is a toxic xenobiotic compound, classified as an endocrine disrupter capable of interfering with living organisms. Directive 2003/53/EC of the European Union (EU) prohibited the use of NPEOs and nonylphenols (NP) and recommended replacing them by alcohol ethoxylates, which are expensive [20]. However, due to their low cost and higher efficiency, NPnEOs are still widely used in China. They are discharged continuously into the environment, and the occurrence of NPnEOs in environmental samples has been well documented [21].

In this study, we presented a low-GNP with varying glycine/nitrogen (G/N) ratios for synthesizing Cu_xO_y and metallic copper catalysts. Furthermore, the degradation of NP9EO under ultraviolet (UV) irradiation was investigated using a heterogeneous Cu_xO_y photocatalyst (and metallic copper) in a custom-made photoreactor. Moreover, the effects of operational parameters, such as catalyst activity, catalyst dosage, and H_2O_2 concentration, were examined. This study will provide useful evidence on the potential of this method as a treatment to remove toxic materials from the water environment.

2. Results and Discussion

2.1. Characterization of Cu-Based Nanoparticles

Figure 1 depicts the X-ray powder diffraction (XRD) pattern of as-synthesized copper-based nanoparticles synthesized through the GNP with different G/N ratios. The XRD pattern of the sample

(a) obtained at a stoichiometric G/N ratio of 0.3 displays peaks at 32.5°, 35.7°, 38.7°, 48.5°, 53.6°, 58.3°, 61.6°, 66.3°, and 68.1° for (110), (−111), (111), (−202), (020), (113), (022) (−311), and (220) planes, respectively, confirming the formation of the CuO monoclinic phase (JCPDS File No. 01-089-5897). At a stoichiometric G/N ratio of 0.5 (see sample (b) in Figure 1), intense reflections were observed at 2θ values of 29.5°, 36.8°, 38.7°, 42.3°, 43.4°, 50.5°, 61.4°, and 74.2°. The peaks at 38.7° corresponded to the (111) plane of divalent CuO, whereas the reflections at 29.5°, 36.8°, 42.3°, and 61.4° corresponded to the (110), (111), (200), and (220) planes of cuprite, respectively (JCPDS File No. 01-078-0428), which indicated Cu_2O nanocrystal formation (JCPDS File No. 01-077-0199) [22]. In addition, the peaks at 50.5° and 74.2°, which corresponded to the (200) and (220) peaks, respectively, of zero-valent copper (Cu^0) were observed. These results clearly indicated that the Cu^0 nanoparticles, formed during chemical reduction, underwent oxidation because of the limited stability of Cu [23,24], and Cu_2O might have been formed through oxidation [25]. When the G/N ratio increased to 0.7, broad reflections from a highly dispersed Cu^0 phase were observed, as shown in sample (c) in Figure 1. The XRD patterns (111), (200), and (220) of Cu^0 planes at 43.4°, 50.5°, and 74.2°, respectively, were consistent with bulk copper crystallographic data (JCPDS File No. 01-085-1326). The results clearly revealed that Cu^0 formation was favored with an increase in glycine (fuel) content during combustion synthesis. When the G/N ratio was reduced to 0.3, sample (a) exhibited numerous primary CuO phases and a few Cu_2O phases (approximately 9.5%). These findings indicated that nanocrystalline CuO reduced to a stable Cu_2O phase rather than forming metallic copper directly [26]. The average crystallite size was calculated from the XRD pattern according to the line width of the (200) plane refraction peak by using the Debye–Scherrer equation [27]:

$$d = \frac{\kappa_1 \times \lambda}{\beta \times \cos\theta} \quad (1)$$

where d is the crystallite size (nm), k_1 is the constant with a value of approximately 0.9, λ is the wavelength of the X-ray radiation ($\lambda = 0.1542$ Å CuKα), β is the full width at half maximum intensity, and the value of 2θ is the characteristic peak of the XRD pattern. The crystalline sizes of copper-based nanoparticles were 76.3 ± 2.2 nm (G/N = 0.3), 35.4 ± 1.0 nm (G/N = 0.5), and 99.4 ± 2.9 nm (G/N = 0.7), respectively, as presented in Table 1.

Figure 2 shows the SEM with EDS images of as-prepared Cu-based nanoparticles at various G/N ratios. The typical EDS spectra of the particles (Figure 2) indicated that the particles were composed of copper (Cu) and oxygen (O). The SEM photographs showed that the catalysts formed sponge-like aggregates containing some nanoparticles; the size distribution range of the synthesized particles was broad (5–100 nm), and some irregular granules with a high degree of shape anisotropy were observed. The prepared Cu-based nanoparticles were not single-domain particles. The lack of homogeneity in particle shape and size, as well as the presence of sponges or flake-like structures, was typical of combustion synthesis. The results indicated that the large amounts of gas that evolved during combustion affected the porosity of the solid products [28,29]. Moreover, high combustion temperatures promoted nanoparticle agglomeration and crystal grain growth [30]. This finding was consistent with the surface area (BET) analysis results in the present study (Table 1). The BET results indicated that increasing the glycine concentration and reducing the particle surface area promoted the formation of metallic Cu. Consequently, large-scale agglomerates were observed when the ratio G/N = 0.7 was used.

Table 1. Specific surface area, crystallite size, and determination of copper oxide nanoparticles synthesized at different glycine/nitrogen (G/N) ratios.

Catalyst	Synthesis of Copper Oxide Nanoparticles Ratio Analysis [%]			Specific Surface Area ($m^2\ g^{-1}$)	Crystallite Size (nm)
	CuO	Cu_2O	Cu		
G/N = 0.3	90.5	9.5	0	166.9	76.3
G/N = 0.5	43.2	54.6	2.2	131.0	35.4
G/N = 0.7	0	24.8	75.2	92.1	99.4

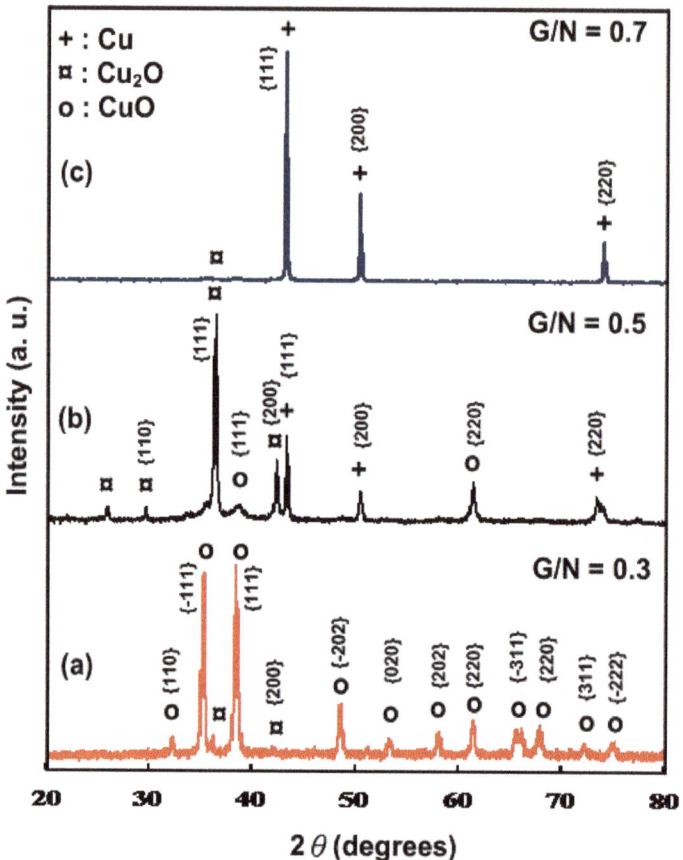

Figure 1. The X-ray diffraction (XRD) pattern of combustion synthesized products of copper-based materials with various glycine/nitrate ratios (G/N) = 0.3 (**a**), 0.5 (**b**), and 0.7 (**c**).

X-ray photoelectron spectroscopy (XPS) spectra of the selected catalysts were obtained to confirm the surface composition and oxidation state of copper from information derived from binding energy values. Copper oxidation states occurred during two semiconducting phases: cupric oxide (CuO) and cuprous oxide (Cu_2O) [31]. The binding energies in the XPS analyses were calibrated using the C 1s peak (BE = 284.6 eV), where the C 1s peak was attributed to residual carbon from the sample and adventitious hydrocarbons from the XPS instrument or adsorbed carbon. Figure 3 presents the three samples obtained from the XPS results. As displayed in Figure 3a, the main O 1s peak present at 532.8 eV was attributed to lattice oxygen in a metal oxide, such as CuO or Cu_2O. The deconvolution of the Cu 2p3/2 XPS core level spectra of Cu_xO_y is illustrated in Figure 3b–d. Through spectral deconvolution, we observed that the binding energies of Cu_2O and CuO were in the ranges 933.1–933.7 and 934.3–934.5 eV for different samples, respectively [32,33]. In addition, two peaks were observed, and the feature at 932.4 ± 0.1 eV was attributed to metallic Cu (2p3/2) [31,32], as presented in Figure 3c,d, respectively.

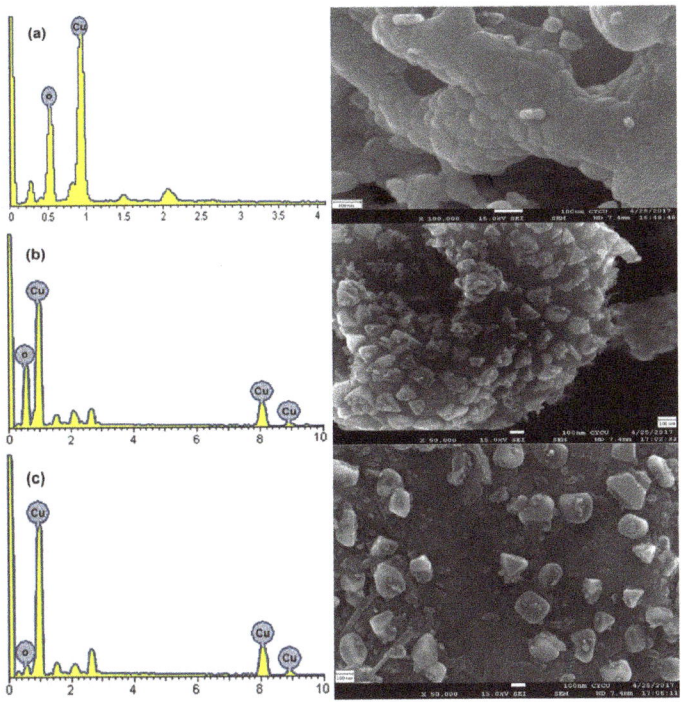

Figure 2. SEM with EDS images of prepared Cu-based nanoparticles at various G/N ratios: G/N = 0.3 (**a**), G/N = 0.5 (**b**), and G/N = 0.7 (**c**).

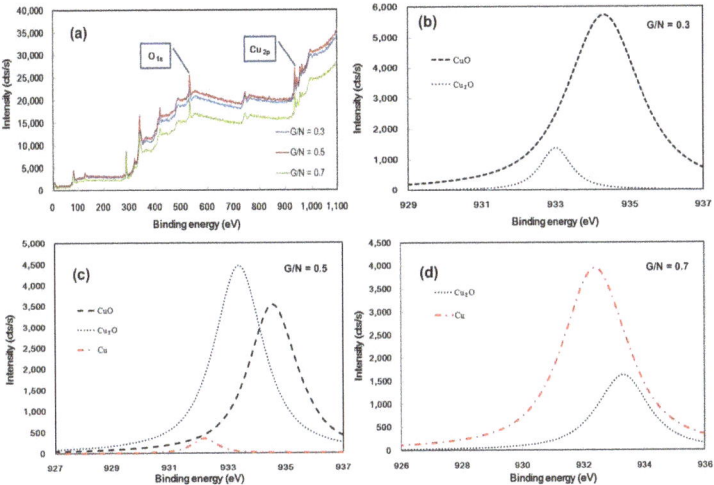

Figure 3. X-ray photoelectron spectroscopy (XPS) wide-scan spectra (**a**) of all samples; deconvolution of XPS spectra of (**b**) (G/N) = 0.3, (**c**) (G/N) = 0.5, and (**d**) (G/N) = 0.7 samples.

2.2. Investigation of Photocatalytic Performance

NP9EO degradation was used in this study to evaluate the photocatalytic activity of the as-prepared samples; its chemical structure is demonstrated in Figure 4. Economically, catalyst dosage is a main parameter in degradation studies. As illustrated in Figure 5, the custom-made photoreactor (Panchum Scientific Co., PR-2000) was operated in the atmosphere and 14 UV light tubes functioned as germicidal lamps with a wavelength of 254 nm (Sankyo Denki Co., Ltd., Kanagawa, Japan), surrounding the cylindrical quartz reactor in sequence. Of these, a light with a power of approximately 10 mW/cm^2 was placed in the center of the reactor and measured by a power meter (Model 840-C Handheld). To optimize the dosage of nano-Cu for photocatalyst activity, experiments were performed by varying the nano-Cu (G/N = 0.3) amount from 0.1 to 1 g/L. Figure 6a illustrates that the degradation efficiencies for 0.1, 0.3, 0.5, 0.8, and 1.0 g/L nano-Cu (G/N = 0.3) were 80.1%, 83.1%, 78.2%, 73.6%, and 70.3%, respectively. Besides, the TOC value represents the total organic carbons concentration in the solution. It is an important indicator of mineralization [34]. Figure 6a also indicated that TOC removal efficiency and NP9EO degradation efficiency showed the same trend. TOC removal efficiency increased from 58.2% for a nano-Cu catalyst amount of 0.1 g to more than 70.5% for a nano-Cu catalyst amount of 0.3 g. However, the concentration of nano-Cu catalysts (G/N = 0.3) for 0.5 g/L was increased above the limiting value, and TOC removal efficiency decreased (the mineralization efficiency dropped from 60.7% to 54.9%) because of increasing suspension turbidity and decreasing light penetration from an increased scattering effect [35], as shown in Figure 6a. Therefore, the optimal dosage of nano-Cu catalysts (G/N = 0.3) in this custom-made photoreactor was 0.3 g/L in the present study.

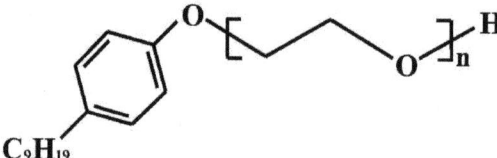

Figure 4. Chemical structure of NPnEO.

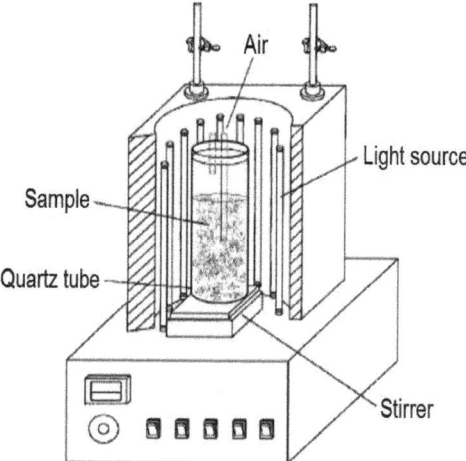

Figure 5. Schematic representation of the custom-made photoreactor.

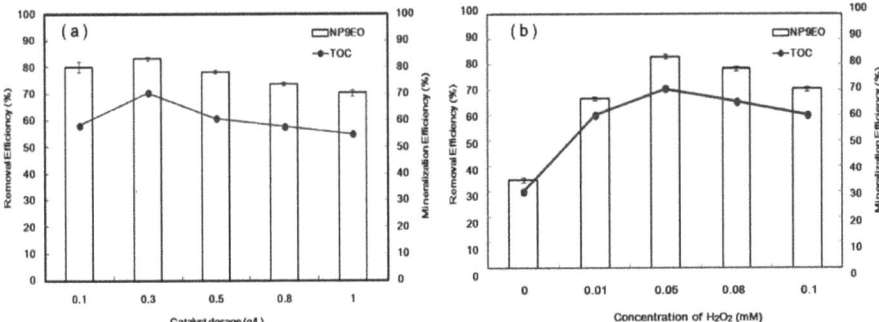

Figure 6. (a) Effect of catalyst dosage on the degradation of NP9EO (G/N = 0.3, NP9EO = 100 mg/L, H_2O_2 = 0.05 mM (1000 mg/L), pH = 6), (b) Effect of increasing H_2O_2 concentration on NP9EO degradation. Data presented are the average of duplicate experiments with relative standard deviation. (G/N = 0.3, NP9EO = 100 mg/L, pH = 6, dosage = 0.3 g/L).

In addition, the effects of hydrogen peroxide (H_2O_2) concentration (0–0.1 M) on photocatalytic oxidation were investigated. The experimental results indicated that, initially, on increasing the H_2O_2 concentration up to 0.05 mM for nano-Cu catalysts, the highest exhibited NP9EO and TOC removal efficiencies were 83.1% and 70.6%, respectively; the rate of degradation increased due to the availability of the hydroxyl radical (·OH) by the decomposition of more H_2O_2 molecules [36]. Subsequently, NP9EO and TOC removal efficiency gradually decreased (70.3% and 60.1%, respectively) with increasing H_2O_2 concentration above 0.08 mM (Figure 6b). The propagation step in the oxidative cycle is hindered by an excess of H_2O_2, which scavenges the ·OH radicals in the solution due to the reaction between excess H_2O_2 and ·OH radicals (Reactions (2) and (3)), thereby reducing the amount of ·OH available for reaction with NP9EO [37,38].

$$H_2O_2 + OH \rightarrow HO_2 + H_2O \tag{2}$$

$$HO_2 + HO_2 \cdot \rightarrow O_2 + H_2O \tag{3}$$

The effect of different G/N ratios of Cu-based nanoparticles on photocatalytic activity was also investigated in this study (Figure 7). For photocatalytic activity, each respective dose of Cu-based nanoparticles was dispersed in 1L NP9EO aqueous solution, and then the suspension was stirred in a dark condition for 30 min to establish the adsorption–desorption equilibrium. Subsequently, the suspension was subjected to UV light irradiation. In addition, the background experiments, such as NP9EO, were performed to ensure that the actual photocatalytic activity came from the photocatalysts without the influence of direct pollutant degradation by light irradiation. Photolysis experiment was conductedin the reactor without photocatalysts at 254 nm UV light irradiation. The results in Figure 7 clearly indicate that no significant NP9EO photolysis was detected under UV light irradiation. The order of photocatalytic activity was as follows: G/N = 0.5 > G/N = 0.7 > G/N = 0.3. It was observed that the percentage of NP9EO degradations with different G/N ratios were 83.1%, 94.6% and 87.7%, respectively, and the corresponding TOC removal percentages were 52.8%, 70.6%, and 57.6%, respectively. These results clearly suggested that at G/N = 0.5, the catalyst exhibited a high activity, which was represented as the degradation of NP9EO. This finding is consistent with the results of the XPS analysis of the oxidation state of copper (see Table 1). Cu_2O (bandgap of approximately 2.2 eV) has a higher electric conductivity than CuO (bandgap of approximately 1.6 eV) [39]; therefore, in the experiment on photocatalytic performance, Cu_2O exhibited a higher removal effect than CuO under the UV-light irradiation system in our study. Furthermore, Cu_2O and CuO are more stable at a higher potential for photoelectrochemical/hydrogen-evolution reaction because a high potential restrains the photoreduction of Cu_2O (or CuO) to metallic copper [40]. Dasineh et al. [39] reported that the integration of Cu_2O and CuO significantly increased charge collection and reduced the recombination rate inside the photocatalyst. Luna et al. [41] also reported that the Cu species modified catalysts presented a high photocatalytic activity which was attributed to an appropriate amount of copper phases acting as photogenerated electron traps, avoiding charge recombination. This indicates that Cu_2O and CuO improved optical absorption and facilitated charge transfer at the interface between the photocatalyst and electrolyte. Consequently, with copper-based catalysts in a heterogeneous system, the highest photocatalytic performance was obtained with a G/N ratio of 0.5. In our results, the efficiency of TOC removal was consistently lower than that of NP9EO removal, indicating that, in addition to CO_2 (formed by mineralization), some intermediates were generated (such as by-products) [42].

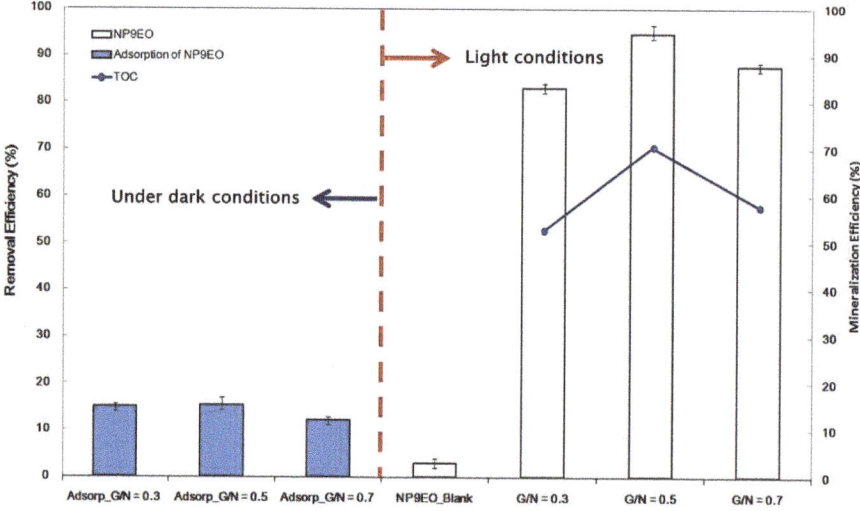

Figure 7. NP9EO degradation curves for Cu-based catalysts synthesized at different G/N ratios. Data presented are the average of duplicate experiments with relative standard deviation. (Catalyst dosage = 0.3g/L, NP9EO = 100 mg/L, H_2O_2 = 0.05 mM (1000 mg/L), pH = 6.)

2.3. Analysis of Variance (ANOVA) Study of Photocatalytic Activity

We further evaluated the effects of nano-Cu_xO_y catalyst dosage (range 0.1–1 g/L), H_2O_2 concentration (range 0–0.1 mM), and different G/N ratios (range 0.3–0.7) on the degradation of NP9EO by using a one-way ANOVA analysis. The nano-Cu_xO_y photocatalytic activity assays were performed in triplicates. For all statistical analyses, values were presented as the mean ± standard deviation, and the significance between different groups was examined ($p < 0.05$). According to the results in Table 2, the nano-Cu_xO_y catalyst had significant effects as the p values were 4.99×10^{-7}, 1.42×10^{-13}, and 9.46×10^{-6} for catalyst dosage, H_2O_2 concentration, and different G/N ratio percentage removals, respectively. This suggests that there were significant differences in the effect of the nano-Cu_xO_y catalyst on the percentage removal of dosage, H_2O_2 concentration, and different G/N ratios, hence this practice can be effectively used to explain the effect of nano-Cu_xO_y on photocatalytic activity efficiency.

Table 2. One-way ANOVA results of various Cu-based nanoparticles effects on the degradation of NP9EO.

	Source of Variation	Sum of Squares	Df	Mean Square	F Value	p Value	Decision
Catalyst dosage	Between groups	329.96	4	82.49	62.24	4.99×10^{-7}	Significant
	Within groups	13.25	10	1.32			
	Total	343.22	14				
H_2O_2 concentration	Between groups	4314.67	4	1078.67	1325.14	1.42×10^{-13}	Significant
	Within groups	8.14	10	0.81			
	Total	4322.81	14				
G/N ratios	Between groups	199.64	2	99.82	62.83	9.46×10^{-6}	Significant
	Within groups	9.53	6	1.58			
	Total	209.18	8				

3. Materials and Methods

3.1. Catalyst Preparation

Cu-based materials were synthesized through the GNP. Copper nitrate trihydrate [$Cu(NO_3)_2 \cdot 3H_2O$] (Merck, Kenilworth, NJ, USA, 99.5% purity) and glycine [$C_2H_5NO_2$] (Merck, Kenilworth, NJ, USA, 99.7% purity) were dissolved in deionized water. Various amounts of glycine (fuel) were added to the metal nitrate (oxidizer) solution to obtain three different glycine–nitrate (G/N) ratios (G/N = 0.3, 0.5, and 0.7). As shown in Figure 8, the glycine–nitrate solutions were then mixed and heated overnight over a hot-plate stirrer (Corning, Glendale, AZ, USA) at 105 °C to form clear, homogeneous, and viscous gel-like solutions. Each gel was then poured into a ceramic bowl that was placed in an oven. The gels were heated to 200 °C until they underwent self-ignition and produced catalyst ashes. Finally, we obtained various shades, from russet brown to dark brown, of copper oxide nanoparticles displayed in the forms of fine powder.

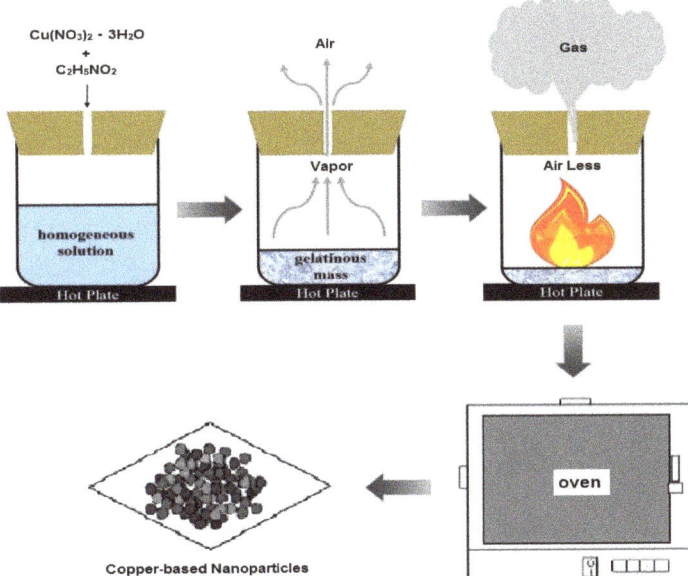

Figure 8. Schematic illustration of combustion synthesis for the copper G/N reaction system.

3.2. Characterization

The crystal structure of the samples was determined through XRD by using Cu Kα radiation ($\lambda = 1.5418$ Å) on a Rigaku DMAX 2200VK/PC diffractometer (Tokyo, Japan). The allocation/distribution of all measured peaks from XRD referred to the data from the Joint Committee on Powder Diffraction Standards (JCPDS) [43]. The XPS (VG Escalab 250 iXL ESCA, VG Scientific, West Sussex, UK) was applied to verify the chemical composition of the samples to obtain the chemical analysis. The specific surface area ($m^2\ g^{-1}$) was calculated by the BET equation [44], and the total pore volume (Vt, $m^3\ g^{-1}$) was examined by converting the adsorption amount at $P/P_0 = 0.95$ to the volume of the liquid adsorbate. The morphologies of the synthesized samples were analyzed through SEM (S-4800, Hitachi, Tokyo, Japan), coupled with EDS measurements.

3.3. Photocatalytic Experiments

The stock solutions of NP9EO were prepared in distilled water to obtain a concentration of 100 mg/L, which was then ultrasonicated for 30 min to obtain a stable emulsion. The emulsion was stored at 4 °C for a maximum of 4 weeks. The photocatalytic activities of nano-Cu samples under UV irradiation were evaluated based on the degradation rate of NP9EO in a quartz reactor (120/115 mm outer/inner diameter, height: 240 mm) containing 0.3 g of nano-Cu and 1 L of a 100 mg/L aqueous solution of NP9EO. The system was operated in the atmosphere and 14 UV light tubes functioned as germicidal lamps with a wavelength of 254 nm (Sankyo Denki Co., Ltd., Kanagawa, Japan), surrounding the cylindrical quartz reactor in sequence.

3.4. Analytical Procedures

The photo-degradation rate of NP9EO solution was determined through HPLC (Agilent 1100 series; Agilent Technologies, Santa Clara, CA, USA) with a Wondasil-C18 (4.6 × 150 mm, 5 μm), and a 2487-UV detector at 277 nm. TOC was measured by a model 1010 TOC analyzer (O.I. Analytical, College Station, TX, USA) following the NIEA W532.52C standard methods (Taiwan's Environmental Protection Agency). Of these, the principle of TOC was calculated by the subtraction of the IC (inorganic carbon) value from the TC (total carbon) of the sample. The pH value was measured with a pH meter (Model TS-1, Suntex Co., New Taipei City, Taiwan).

4. Conclusions

A nanocrystalline Cu-based heterogeneous catalyst was synthesized using the GNP. The nature and amount of fuel were crucial factors controlling combustion and final product composition. The formation of CuO, mixed-valence copper oxides (CuO + Cu_2O), and metallic Cu was achieved by varying the G/N ratio, in which glycine was used as fuel. The effect of the G/N ratio on catalyst morphology was analyzed using SEM. With a G/N ratio of 0.7, agglomerate structure formation was observed. Furthermore, Cu-based catalysts exhibited high photocatalytic activity during NP9EO degradation in a custom-made photoreactor under UV light. The experimental results revealed that, under optimal reaction conditions, namely a catalyst dosage of 0.3 g/L and H_2O_2 concentration of 0.05 mM, NP9EO and TOC removal efficiency was 83.1% and 70.6%, respectively. However, the concentration of nano-Cu catalysts (G/N = 0.3) for 0.5 g/L was increased above the limiting value, and TOC removal efficiency decreased because of increasing suspension turbidity and decreasing light penetration from an increased scattering effect. The NP9EO removal efficiency increased with H_2O_2 concentration but decreased when H_2O_2 concentration exceeded 0.05 mM. With Cu-based catalysts in a heterogeneous system, the highest photocatalytic activity was observed with a G/N ratio of 0.5. Furthermore, TOC was consistently lower than the removal efficiency of NP9EO, indicating that, in addition to the mineralization of CO_2, some intermediates were also generated.

Author Contributions: S.-S.C. and H.-H.C. conceived and designed the experiments that produced/generated this paper. S.-S.C. proposed methodology and project administration related to the research areas. H.-H.C. wrote original manuscript, drew figures, reviewed and edited the article. H.-M.L. analyzed quantitative data, statistically analyzed data, reviewed original manuscript, and provided critical comments. L.-W.J. performed the experiments, collected and analyzed the data. S.-Y.C. performed data processing of the experiments, statistically analyzed, reviewed original manuscript, revised grammar and provided critical comments. H.-H.C. and S.-Y.C. contributed to the final revision of the paper. All authors contributed to the manuscript and discussed the results. All authors have read and agreed to the published version of the manuscript.

Funding: This study was supported by a grant from Hsuteng Consulting International Co., Ltd. (No. HCI 20190113-A).

Acknowledgments: The authors gratefully acknowledge the experimental apparatus support from Institute of Environment Engineering and Management, National Taipei University of Technology.

Conflicts of Interest: The authors declare no conflict of interest.

References

1. Theivasanthi, T.; Alagar, M. Nano sized copper particles by electrolytic synthesis and characterizations. *Int. J. Phys. Sci.* **2011**, *6*, 3662–3671.
2. Chick, L.A.; Pederson, L.R.; Maupin, G.D.; Bates, J.L.; Thomas, L.E.; Exarhos, G.J. Glycine-nitrate combustion synthesis of oxide ceramic powders. *Mater. Lett.* **1990**, *10*, 6–12. [CrossRef]
3. Wan, L.; Cheng, J.G.; Fan, Y.M.; Liu, Y.; Zheng, Z.J. Preparation and properties of superfine W–20Cu powders by a novel chemical method. *Mater. Des.* **2013**, *51*, 136–140. [CrossRef]
4. Coutinho, J.P.; Silva, M.C.; Meneghetti, S.M.P.; Leal, E.; de Melo Costa, A.C.F.; de Freitas, N.L. Combustion synthesis of $ZnAl_2O_4$ catalyst using glycine as fuel for the esterification and transesterification of soybean oil: Influence the form of heating. *Mater. Sci. Forum* **2012**, *727–728*, 1323–1328. [CrossRef]
5. Lim, H.H.; Chua, P.N.; Mun, H.P.; Horri, B.A. Synthesis and characterisation of CuO/HNT nano-particles through in-situ glycine nitrate process. *Int. J. Adv. Sci. Eng. Inf. Technol.* **2018**, *6*, 2321–8991.
6. Tajuddin, M.M.; Patulla, M.H.; Ideris, A.; Ismail, M. Self-combustion synthesis of Ni catalysts modified with La and Ce using Glycine–NitrateProcess (GNP). *Malay. J. Catal.* **2017**, *2*, 8–11.
7. Jadhav, L.D.; Patil, S.P.; Chavan, A.U.; Jamale, A.P.; Puri, V.R. Solution combustion synthesis of Cu nanoparticles: A role of oxidant-to-fuel ratio. *Micro Nano Lett.* **2011**, *6*, 812–815. [CrossRef]
8. Laguna-Marco, M.A.; Haskel, D.; Souza-Neto, N.; Lang, J.C.; Krishnamurthy, V.V.; Chikara, S.; Cao, G.; van Veenendaal, M. Orbital magnetism and spin-orbit effects in the electronic structure of $BaIrO_3$. *Phys. Rev. Lett.* **2010**, *105*, 216407. [CrossRef]
9. Jiao, F.; Frei, H. Nanostructured cobalt and manganese oxide clusters as efficient water oxidation catalysts. *Energy Environ. Sci.* **2010**, *3*, 1018–1027. [CrossRef]
10. Tran, T.H.; Nguyen, V.T. Copper oxide nanomaterials prepared by solution methods, some properties, and potential applications: A brief review. *Int. Sch. Res. Not.* **2014**, *2014*, 1–14. [CrossRef]
11. Li, W.Y.; Xu, L.N.; Chen, J. Co_3O_4 nanomaterials in lithium-ion batteries and gas sensors. *Adv. Funct. Mater.* **2005**, *15*, 851–857. [CrossRef]
12. Clarizia, L.; Spasiano, D.; Di Somma, I.; Marotta, R.; Andreozzi, R.; Dionysiou, D.D. Copper modified-TiO_2 catalysts for hydrogen generation through photoreforming of organics. A short review. *Int. J. Hydrog. Energy* **2014**, *39*, 16812–16831. [CrossRef]
13. Janczarek, M.; Kowalska, E. On the origin of enhanced photocatalytic activity of copper-modified titania in the oxidative reaction systems. *Catalysts* **2017**, *7*, 317. [CrossRef]
14. Bandara, J.; Udawatta, C.P.K.; Rajapakse, C.S.K. Highly stable CuO incorporated TiO_2 catalyst for photocatalytic hydrogen production from H_2O. *Photochem. Photobiol. Sci.* **2005**, *4*, 857–861. [CrossRef]
15. Abd-Elkader, O.H.; Deraz, N. Synthesis and characterization of new copper based nanocomposite. *Int. J. Adv. Sci. Eng. Int. J. Electrochem. Sci.* **2013**, *8*, 8614–8622.
16. Toe, C.Y.; Zheng, Z.; Wu, H.; Scott, J.; Amal, R.; Ng, Y.H. Photocorrosion of cuprous oxide in hydrogen production: Rationalising self-oxidation or self-reduction. *Angew. Chem. Int. Ed.* **2018**, *57*, 13613–13617. [CrossRef] [PubMed]
17. Fornasiero, P.; Christoforidis, K.C. Photocatalysis for hydrogen production and CO_2 reduction: The case of copper-catalysts. *ChemCatChem* **2018**, *11*, 368–382.
18. Wang, L.; Zhao, J.; Liu, H.; Huang, J. Design, modification and application of semiconductor photocatalysts. *J. Taiwan Inst. Chem.* **2018**, *93*, 590–602. [CrossRef]

19. Puteh, M.H.; Stuckey, D.C.; Othman, M.H.D. Direct measurement of anaerobic biodegradability of nonylphenol ethoxylates (NPEOs). *Int. J. Environ. Sci. Dev.* **2015**, *6*, 660–663. [CrossRef]
20. Cox, P.; Drys, G. Directive 2003/53/EC of the European Parliament and of the council. *Off. J. Eur. Comm.* **2003**, *178*, 24–27.
21. Ashar, A.; Iqbal, M.; Bhatti, I.A.; Ahmad, M.Z.; Qureshi, K.; Nisar, J.; Bukhari, I.H. Synthesis, characterization and photocatalytic activity of ZnO flower and pseudo-sphere: Nonylphenol ethoxylate degradation under UV and solar irradiation. *J. Alloys Compd.* **2016**, *678*, 126–136. [CrossRef]
22. Karthikeyan, S.; Kumar, S.; Durndell, L.J.; Isaacs, M.A.; Parlett, C.M.A.; Coulson, B.; Douthwaite, R.E.; Wilson, K.; Lee, A.F. Size-Dependent Visible Light Photocatalytic Performance of Cu_2O Nanocubes. *ChemCatChem* **2018**, *10*, 3554–3563. [CrossRef]
23. Aslam, M.; Gopakumar, G.; Shoba, T.L.; Mulla, I.S.; Vijayamohanan, K.; Kulkarni, S.K.; Urban, J.; Vogel, W. Formation of Cu and Cu_2O nanoparticles by variation of the surface ligand: Preparation, structure, and insulating-to-metallic transition. *J. Coll. Interf. Sci.* **2002**, *255*, 79–90. [CrossRef] [PubMed]
24. Khan, A.; Rashid, A.; Younas, R.; Chong, R. A chemical reduction approach to the synthesis of copper nanoparticles. *Int. Nano Lett.* **2016**, *6*, 21–26. [CrossRef]
25. Feng, L.; Zhang, C.; Gao, G.; Cui, D. Facile synthesis of hollow Cu_2O octahedral and spherical nanocrystals and their morphology-dependent photocatalytic properties. *Nanoscale Res. Lett.* **2012**, *7*, 276–286. [CrossRef]
26. Patterson, A.L. The scherrer formula for X-Ray particle size determination. *Phys. Rev.* **1939**, *56*, 978–982. [CrossRef]
27. Pike, J.; Chan, S.W.; Zhang, F.; Wang, X. Formation of stable Cu_2O from reduction of CuO nanoparticles. *Appl. Catal. A Gen.* **2006**, *303*, 273–277. [CrossRef]
28. Patil, K.C.; Hegde, M.S.; Rattan, T.; Aruna, S.T. *Chemistru of Nanocrystalline Oxide Materials: Combustion Synthesis, Properties and Applications*; World Scientific Publishing Co. Pte. Ltd.: Singapore, 2008.
29. Kumar, A.; Cross, A.; Manukyan, K.; Bhosale, R.R.; van den Broeke, L.J.P.; Miller, J.T.; Mukasyan, A.S.; Wolf, E.E. Combustion synthesis of copper–nickel catalysts for hydrogen production from ethanol. *Chem. Eng. J.* **2015**, *278*, 46–54. [CrossRef]
30. Podbolotova, K.B.; Khorta, A.A.; Tarasovb, A.B.; Trusovc, G.V.; Roslyakovd, S.I.; Mukasyan, A.S. Solution combustion synthesis of copper nanopowders: The fuel effect. *Combust. Sci. Technol.* **2017**, *189*, 1878–1890. [CrossRef]
31. Korzhavyi, P.A.; Johansson, B. *Literature Review on the Properties of Cuprous Oxide Cu_2O and the Process of Copper Oxidation*; SKB Report TR-11-08; Swedish Nuclear Fuel and Waste Management Company: Stockholm, Sweden, 2011.
32. Biesinger, M.C. Advanced analysis of copper X-ray photoelectron spectra. *Surf. Interface Anal.* **2017**, *49*, 1325–1334. [CrossRef]
33. Monte, M.; Munuera, G.; Costa, D.; Conesa, J.C.; Martínez-Arias, A. Near-ambient XPS characterization of interfacial copper species in ceria-supported copper catalysts. *Phys. Chem. Chem. Phys.* **2015**, *17*, 29995–30004. [CrossRef] [PubMed]
34. Shu, H.Y.; Chang, M.C.; Tseng, T.H. Solar and Visible Light Illumination on Immobilized Nano Zinc Oxide for the Degradation and Mineralization of Orange G in Wastewater. *Catalysis* **2017**, *7*, 164–179.
35. Siah, W.R.; Lintang, H.O.; Shamsuddin, M.; Yoshida, H.; Yuliati, L. Masking effect of copper oxides photodeposited on titanium dioxide: Exploring UV, visible, and solar light activity. *Catal. Sci. Technol.* **2016**, *6*, 5079–5087. [CrossRef]
36. Zhang, L.; Wang, L.; Liu, P.; Fu, B.; Hwang, J.; Chen, S. Analysis on deep treatment effect of coking wastewater using 3D electrode combined with Fenton reagent. *Charact. Mater. Met. Mater.* **2015**, *1*, 185–192.
37. Nezamzadeh-Ejhieh, A.; Salimi, Z. Heterogeneous photodegradation catalysis of o-phenylenediamine using CuO/X zeolite. *Appl. Catal. A Gen.* **2010**, *390*, 110–118. [CrossRef]
38. Inchaurrondo, N.; Cechini, J.; Font, J.; Haure, P. Strategies for enhanced CWPO of phenol solutions. *Appl. Catal. B Environ.* **2012**, *111–112*, 641–648. [CrossRef]
39. Dasineh Khiavi, N.; Katal, R.; Kholghi Eshkalak, S.; Masudy-Panah, S.; Ramakrishna, S.; Jiangyong, H. Visible light driven heterojunction photocatalyst of CuO-Cu_2O thin films for photocatalytic degradation of organic pollutants. *Nanomaterials* **2019**, *9*, 1011. [CrossRef]
40. Yang, Y.; Xu, D.; Wu, Q.; Diao, P. Cu_2O/CuO Bilayered Composite as a high-efficiency photocathode for photoelectrochemical hydrogen evolution reaction. *Sci. Rep.* **2016**, *6*, 1–13. [CrossRef]

41. Luna, A.L.; Valenzuela, M.A.; Colbeau-Justin, C.; Vázquez, P.; Rodriguez, J.L.; Avendaño, J.R.; Alfaro, S.; Tirado, S.; Garduno, A.; De la Rosa, J.M. Photocatalytic degradation of gallic acid over CuO-TiO$_2$ composites under UV/Vis LEDs irradiation. *Appl. Catal. A Gen.* **2016**, *521*, 140–148. [CrossRef]
42. Chen, S.S.; Hsu, H.T.; Tsui, H.J.; Chang, Y.M. Removal of nonionic surfactant from electroplating wastewater by fluidized zerovalent iron with two oxidants (H$_2$O$_2$/Na$_2$S$_2$O$_8$). *Desalin. Water Treat.* **2013**, *51*, 1678–1684. [CrossRef]
43. Sarkar, D.K.; Paynter, R.W. One-step deposition process to obtain nanostructured superhydrophobic thin films by galvanic exchange reactions. *J. Adhes. Sci. Technol.* **2013**, *24*, 1181–1189. [CrossRef]
44. Brunauer, S.; Emmett, P.H.; Teller, E. Adsorption of gases in multimolecular layers. *J. Am. Chem. Soc.* **1938**, *60*, 309–319. [CrossRef]

 © 2020 by the authors. Licensee MDPI, Basel, Switzerland. This article is an open access article distributed under the terms and conditions of the Creative Commons Attribution (CC BY) license (http://creativecommons.org/licenses/by/4.0/).

Article

Reductive Dechlorination of Chloroacetamides with NaBH₄ Catalyzed by Zero Valent Iron, ZVI, Nanoparticles in ORMOSIL Matrices Prepared via the Sol-Gel Route

Michael Meistelman [1], Dan Meyerstein [2,3,*], Amos Bardea [4], Ariela Burg [5], Dror Shamir [6] and Yael Albo [1,*]

1. Chemical Engineering Dept. and The Centre for Radical Reactions, Ariel University, Ariel 4070000, Israel; michaelme@ariel.ac.il
2. Department of Chemical Sciences and The Centre for Radical Reactions, Ariel University, Ariel 4070000, Israel
3. Chemistry Dept., Ben-Gurion University, Beer-Sheva 8410501, Israel
4. Faculty of Engineering, Holon Institute of Technology (HIT), Holon 5810201, Israel; amos.bardea@hit.ac.il
5. Department of Chemical Engineering, Shamoon College of Engineering, Beer-Sheva 8410802, Israel; arielab@ac.sce.ac.il
6. Department of Chemistry, Nuclear Research Centre Negev, Beer-Sheva 8419001, Israel; drorshamir@gmail.com
* Correspondence: danm@ariel.ac.il (D.M.); yaelyt@ariel.ac.il (Y.A.)

Received: 5 July 2020; Accepted: 20 August 2020; Published: 1 September 2020

Abstract: The efficient reductive dechlorination, as remediation of dichloroacetamide and monochloroacetamide, toxic and abundant pollutants, using sodium borohydride catalyzed by zero valent iron nanoparticles (ZVI-NPs), entrapped in organically modified hybrid silica matrices prepared via the sol-gel route, ZVI@ORMOSIL, is demonstrated. The results indicate that the extent of the dechlorination reaction depends on the nature of the substrate and on the reaction medium. By varying the amount of catalyst or reductant in the reaction it was possible to obtain conditions for full dechlorination of these pollutants to nontoxic acetamide and acetic acid. A plausible mechanism of the catalytic process is discussed. The present work expands the scope of ZVI-NP catalyzed reduction of polluting compounds, first reports the catalytic parameters of chloroacetamide reduction, and offers additional insight into the heterogeneous catalyst structure of M^0@ORMOSIL sol-gel. The ZVI@ORMOSIL catalyst is ferromagnetic and hence can be recycled easily.

Keywords: catalysis; zero valent iron; de-chlorination; borohydride; sol-gel; ferromagnetic

1. Introduction

Halogenated hydrocarbons are known to be toxic pollutants [1]. More than half a century ago it was proposed to study each halogenated compound independently of its homologue hydrocarbon series [1]. While there are number of halogenated pollutants that are monitored by the health authorities, e.g., halo-acetic acids, others which appear on the EPA's (United States Environmental Protection Agency) contaminant candidate list await proper directives and legislation, e.g., halo-acetamides which are formed during chlorination of drinking water [2]. Halo-organic compounds can be reduced electro-chemically [3,4], photo-chemically [5], radiolytically [6], and by a variety of reducing agents [7]. These processes often require a catalyst [8]. M^0-nano-particles (M^0-NPs), e.g., Ag^0-NPs and Au^0-NPs [9,10], are often used as catalysts for these processes [11–14], thus enabling the reduction of these compounds to their non-toxic or less toxic and environmentally friendly products.

Zero valent iron (ZVI) is a feasible solution to environmental applications due to its reactivity, high abundance, and relatively low cost [15,16]. ZVI has been widely studied as a reducing agent for the remediation treatment of halogenated hydrocarbon pollutants in the environment, both for contaminated soils and for water sources [17,18]. Other applications involve the adsorption of toxic heavy metal cation such as lead and arsenic [19–21] and of organic pollutants such as tetracycline [22], which might reduce the risk of increasing antibiotics resistance. It is also used in Fenton-based oxidations, for example, to remove amoxicillin [23] and to remove TNT from water sources [20].

However, ZVI forms surface oxides/hydroxides in an aerobic environment that dramatically reduce the reaction rates, and therefore, ZVI is not used in batch processes that must be completed on a short timescale [24]. The activity of ZVI can be improved by adding Cu^0 to the system [25,26], probably due to a galvanic effect. It was shown that a weak magnetic field enhances the dehalogenation of tri-haloacetamides, however, mono-chloro-acetamides are not reduced [27]. ZVI can also be used as catalyst. For example, when paired with $NaBH_4$ as a reducing agent, it was shown to optimize the reduction of nitrophenol [28–30].

Recently it was shown that ZVI-NPs entrapped in sol-gel matrices can be used as effective heterogeneous catalysts for reductive dehalogenation reactions (RDH) of halo-acetic acids [31]. The high yields and the stability of the catalyst suggest its plausible use for the reduction of the more toxic and very abundant chloroacetamides family of pollutants [2,32–34]. To the best of our knowledge, there has been no attempt to date to utilize ZVI-NPs as a catalyst in the reductive dehalogenation of chloroacetamides. The strategy of degrading dichloroacetamide (DAcAm) in various reaction media has not been reported yet. Shedding light on DAcAm reduction reactivity and the mechanism of reductive dehalogenation reaction in different solvents can help elucidate better ways of disposing of this hazardous pollutant and offer a versatile strategy towards remediation of various environmental pollutants of the same group. In the current study, it is shown that ZVI-NPs entrapped in organically modified silica (ORMOSIL) matrices (ZVI@ORMOSIL), prepared via the sol-gel route, are useful catalysts for the reductive dehalogenation of dichloroacetamide with $NaBH_4$, yielding acetamide and acetate as the final harmless products.

2. Results and Discussion

2.1. 1.0 M ZVI-NPs Ethanolic Suspension

The 1.0 M ZVI-NPs suspension, prepared from the 25 nm ZVI commercial powder, was analyzed using transmission electron microscopy to ensure that no major agglomeration had occurred in the suspension used for preparation of the catalyst (Figure 1). Although agglomerated macro-particles are observed, they largely consist of recognizable nanosized ZVI spheres that fall in the nanometer range, <100 nm. Hence, they are suitable to be used as nano-catalysts. This colloidal suspension was used in the ZVI@ORMOSIL catalyst preparation.

The ethanolic suspension shows magnetic properties that were observed by simple magnet-induced moving of the suspended ZVI powder within the solution.

Photon cross correlation spectroscopy measurements were performed to estimate the initial particle aggregation degree due to interaction of the particles in the suspension. The results obtained support the TEM findings indicating some agglomeration of the ZVI NPs (Figure 2). Although it might be suspected that the agglomerated particles that are observed in Figure 1 might originate from solvent evaporation that occurred while placing the suspension sample on the grid, the average values of Sauter mean diameter (SMD), 95.5 nm, and volume mean diameter (VMD) of 95.5 nm, measured for 25 nm ZVI commercial powder in 1.0 M ethanolic suspension, show that agglomeration of a similar degree occurs in the suspension.

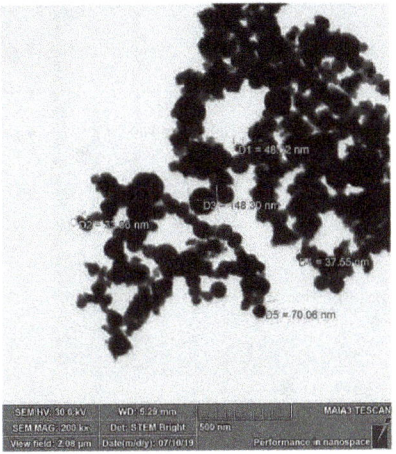

Figure 1. TEM micrograph of 1.0 M ethanolic suspension of commercial 25 nm zero valent iron nanoparticles (ZVI-NPs).

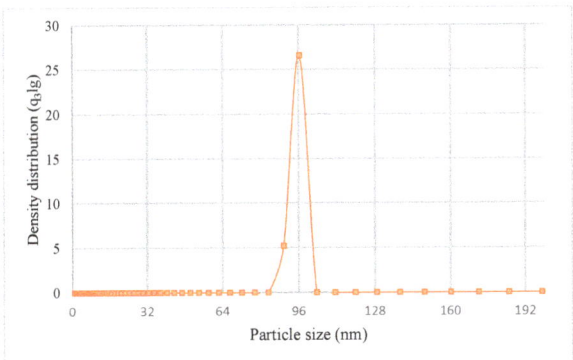

Figure 2. Particle size distribution of 1.0 M ethanolic suspension of commercial 25 nm ZVI-NPs.

ZVI@ORMOSIL Synthesis via the Sol-Gel Route

The differences between ZVI@ORMOSIL gel forms in different preparation steps are visualized in Figure S1. As the wet gel dries in the desiccator, the raw gel undergoes shrinkage due to cooperative action between hydrolysis and densification [35]. In the final stages of the preparation, the dried gel is crushed into powder using a mortar and pestle. The obtained powder has magnetic properties (Figure 3).

Figure 3. Dry grinded ZVI@ORMOSIL powder on a magnetic bar.

2.2. Catalyst Characterization

To investigate the surface morphology and to determine the elemental composition of 1.0% ZVI@ORMOSIL, STEM/EDAX analysis was performed (Figure 4).

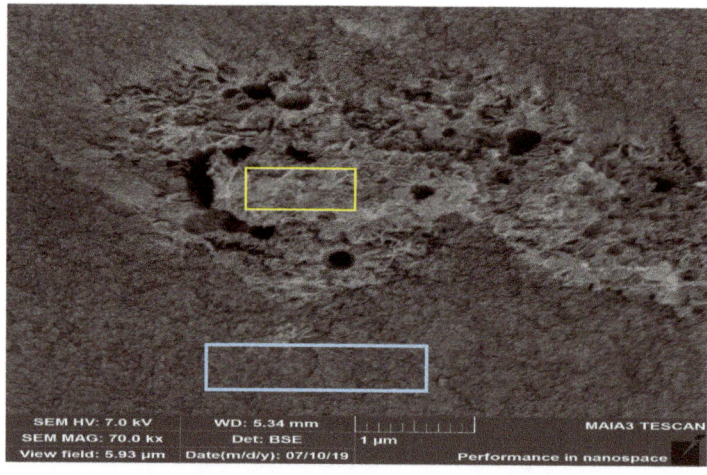

(a)

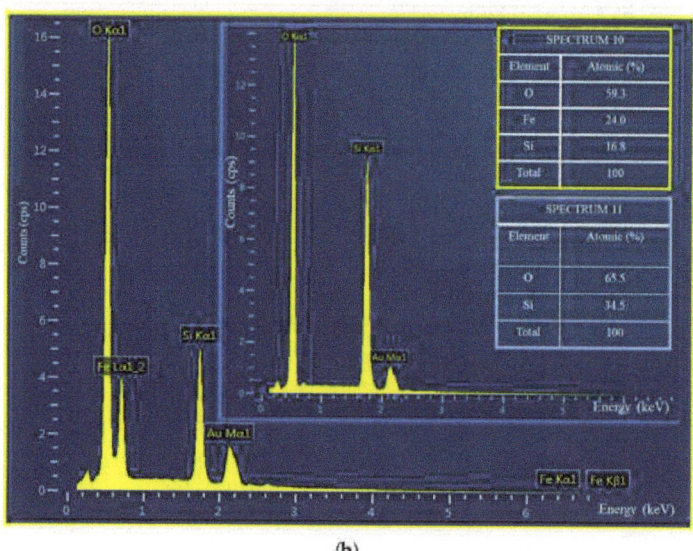

(b)

Figure 4. Scanning electron microscopy image and elemental composition of ZVI@ORMOSIL catalyst matrix. (**a**) SEM image of 1.0% ZVI@ORMOSIL matrix. (**b**) Elemental composition measured by electron dispersive analysis in two areas marked in (**a**); Spectrum 10, measured near cracks and breaks in the matrix-yellow frame; Spectrum 11, measured in the monolith bulk of ZVI@ORMOSIL matrix-blue frame.

The results indicate that the amorphous ORMOSIL surface comprises many cracks, probably due to non-uniform shrinkage and densification of the matrix in the drying stage of the sol-gel process. The elemental composition spectra were taken from two regions of the monolith, marked in Figure 4a.

Near the cracks and breaks formed in the structure, the composition shows the presence of elemental iron (spectrum 10). Conversely, signals collected from the uniform part of the bulk (spectrum 11) exhibit no significant elemental iron. Nevertheless, traces of elemental iron might be present on the monolith surface. It is possible that ZVI will not be detected if it is at a low concentration level and below the sensitivity level of the specific apparatus. A surface composition sensitive method, XPS, was applied to ensure more precise iron detection, as shown below.

Discrimination between successful entrapment in the voids of the bulk material and surface immobilization is possible by measuring the surface composition. To characterize the surface composition of the ZVI@ORMOSIL catalyst, XPS analysis was used. The survey scan of the whole range of 1.0% load ZVI@ORMOSIL is presented in Figure 5. Silicon and oxygen were successfully observed. The carbon detected is probably that of the methyl groups in the matrix originating from the methyltrimethoxysilane precursor. Focused scans for iron content led to detection of almost negligible traces, close to the baseline noise level, as shown in the insert of Figure 5. Using manual sample manipulation, a successful integration extracted a sample area that was calculated as 0.2% atomic of the whole sample area that was scanned in the analysis. Although iron was found as a part of the surface composition, as summarized in Table S1, the almost negligible amounts confirm that most of the catalyst was entrapped within the bulk of the hybrid silica host structure. These findings indicate that the ZVI is trapped in the pores, and they exclude the unwanted possibility of significant surface immobilization of the catalyst.

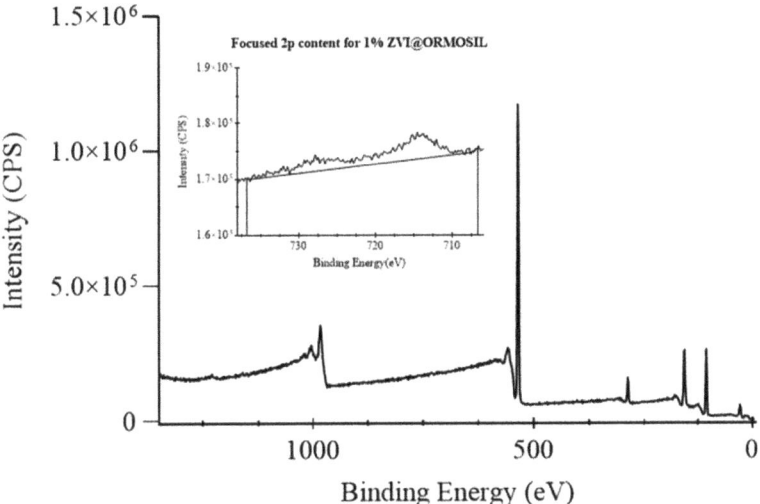

Figure 5. XPS spectra of 1.0% load ZVI@ORMOSIL. Insert: focused scan on Fe^0 peak.

Three types of samples were analyzed using powder X-ray diffraction, PXRD: the original 1.0 M suspension of ZVI used in the preparation of matrices, 1.0% ZVI-NP load ZVI@ORMOSIL, and 10% load ORMOSIL, which was prepared to aid the identification of ZVI phases. Incorporation of low amounts of ZVI nanoparticles in the bulk structure might result in diffraction patterns that are obscured due to the amorphous silica baseline signal. Figure 6 shows a comprehensive overlap among phases found in the original ZVI suspension that was used in the preparation of 1.0% and 10% load ZVI@ORMOSIL, as summarized in Table 1. Major phases were seen at 44.7° and 65.1°, fitting the 1,1,0 and 2,0,0 plains of Fe^0 phases. The slight presence of a Fe_3O_4 phase was also confirmed with peaks at 35.4° and 62.5°. These findings fall in line with the expected observations since an almost neglectable catalyst presence was found in the surface layers of the gel and considerable presence was confirmed for the penetrative XRD analysis of ZVI@ORMOSIL.

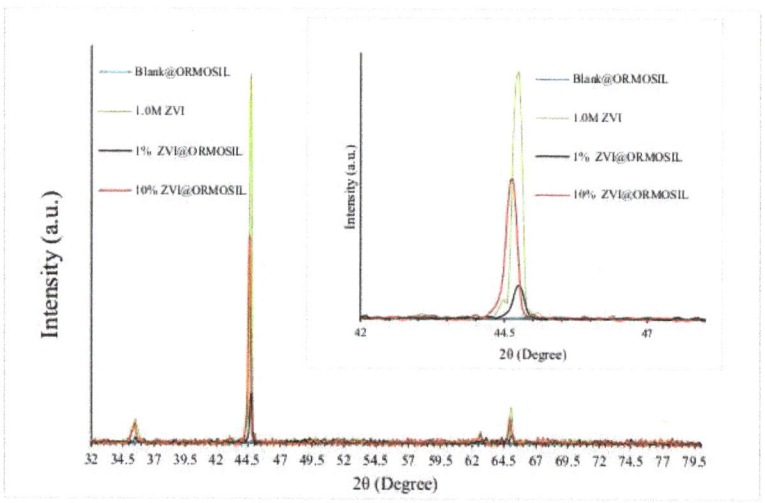

Figure 6. PXRD diffractograms. Blank@ORMOSIL-black, 1.0 M ZVI ethanolic suspension—blue, 10% mol load ZVI@ORMOSIL—green, 1.0% mol load ZVI@ORMOSIL—red.

Table 1. Metal phases found in ZVI@ORMOSIL.

Substance	2θ	Miller Index	File No.
Fe^0	44.7°	1,1,0	04-007-9753
Fe^0	65.1°	2,0,0	
Fe_3O_4	35.4°	3,1,1	04-005-4319
Fe_3O_4	62.5°	4,0,0	

The Scherrer equation was used to calculate the average crystallite size (Dp, nm). Reflection parameters at 44.7° and 65.1° were used for the calculation; the results were averaged and are summarized in Table 2 (all the relevant phases and card numbers files are given in the Supplementary Materials (SI-3)). A certain agglomeration extent might be attributed to the influence of ammonium hydroxide [36] introduced in the preparation of ZVI@ORMOSIL (2%NH_3, 1.5 mL). Similarly prepared catalysts previously reported [31] had an average crystallite size of 34.5 nm, as calculated from the powder XRD results. It was noted that a ZVI secondary particle might comprise many primary crystallites.

Table 2. Average crystallite size.

Material	Dp (nm)
1.0 M susp.	44.9
10% ZVI@ORMOSIL	38.8
1% ZVI@ORMOSIL	34.7

A descriptive ORMOSIL sol-gel pore system and framework study of the catalyst structure was performed using N_2 adsorption-desorption studies. Surface area and pore volumes for 1.0% ZVI@ORMOSIL and for the blank@ORMOSIL matrix without nanoparticles are summarized in Table 3. Both the host matrix without particles and the entrapped catalyst matrix have large surface areas and narrow distributions of the inner pores that are in the mesoporous range. The pore size distributions (diameter) are presented in Figure S4. Surprisingly, the doped matrix has a larger pore volume and surface area; this result suggests that that the agglomeration of the gel is affected by the presence of

ZVI-NPs added during the gelation stage. The larger surface area of ZVI@ORMOSIL induces the reaction substrates to be adsorbed more onto its surface.

Table 3. N_2 adsorption-desorption isotherms summary.

Sample	Surface Area (m²/g)	Average Pore Volume (cm³/g)	Average Pore Diameter (nm)
Blank@ORMOSIL	574	0.53	4.7
1.0% ZVI@ORMOSIL	751	0.73	4.7

The N_2 adsorption-desorption isotherms presented in Figure 7 can be classified as type IV isotherms with H2A hysteresis loops. The asymmetrical shape of these H2A hysteresis loop types (as per current IUPAC convention [37]) are typical for porous glass [38]. These loops describe a distribution of pore sizes and shapes with bottleneck constrictions [39]. They are characteristic of interconnected ink-bottle pores rather than isolated, individual ink-bottle pores, as an "assembly of cavities connected by constrictions" [39]. These findings correspond to previously reported organically modified silica (ORMOSIL) matrices prepared by mixing a large variety of precursors [40,41]. Similar hysteresis loops were previously reported for ORMOSIL entrapped metal nanoparticles used in the reductive dehalogenation of halo-acetic acids [13].

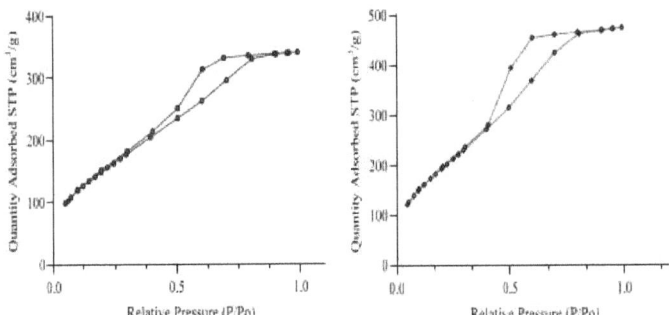

Figure 7. N_2 adsorption-desorption isotherms Blank@ORMOSIL (**left**) and 1.0% ZVI@ORMOSIL (**right**).

2.3. Catalytic Reductive Dehalogenation

2.3.1. Dehalogenation of Di- and Mono-Chloroacetamides

The reductions of monochloroacetamide (MAcAm), dichloroacetamide (DAcAm), and of solutions containing both substrates at a 1:1 molar ratio were performed in different reaction media. The reactions were performed at an initial pH of 8.0 with a constant substrate:$NaBH_4$ molar ratio of 1:20, if not otherwise stated. The products distributions obtained under each reaction condition are summarized in Table 4.

To check the reduction rates of ZVI alone as a reducing agent, typical dehalogenation reactions of both DAcAm and MAcAm were performed, the absence of the sodium borohydride addition step being the only difference. These reactions were continued for 24 h with no reduction yields worth mentioning. Thus, the $NaBH_4$ addition is required. $NaBH_4$ was added to the homogenized suspension of the substrates and catalysts and 2.0 mL of water was added shortly afterward to ensure the full dissolution of $NaBH_4$. The products analysis was performed 15 min later. This time frame choice was designed to assist in the mechanistic study displaying the distribution of the reaction products under various conditions. The plausible reduction products for DAcAm; MAcAm, acetamide (AcAm),

Acetic acid (AA), and ammonia are given in reaction (1) and the product distributions obtained by its reduction are presented in Figure 8.

$$\text{Cl}_2\text{CHC(O)NH}_2 \rightarrow \text{ClCH}_2\text{C(O)NH}_2 \rightarrow \text{CH}_3\text{C(O)NH}_2 \rightarrow \text{CH}_3\text{COOH} + \text{NH}_3 \quad (1)$$

Table 4. The medium effect on dehalogenation yields of DAcAm and MAcAm *.

Starting Subs.	Medium	DAcAm (%)	MAcAm (%)	AcAm (%)	AA (%)
DAcAm	H$_2$O	0	76	18	6
MAcAm		0	53	47	0
DAcAm/MAcAm, (1/1)		0	65	31	3
DAcAm	EtOH:H$_2$O 8:2	22	63	10	5
MAcAm		0	84	16	0
DAcAm/MAcAm, (1/1)		12	75	13	3
DAcAm	ACN:H$_2$O 8:2	46	52	2	0
MAcAm		0	80	20	0
DAcAm/MAcAm, (1/1)		23	66	11	0
DAcAm	2-PrOH:H$_2$O 8:2	60	40	0	0
MAcAm		0	90	10	0
DAcAm/MAcAm, (1/1)		30	65	5	0

* 8.8 mM and 4.4 mM DAcAm and MAcAm were reduced with [NaBH$_4$] = 0.176 M, 0.10 g ZVI@ORMOSIL, for 15 min at room temperature, initial pH 8.0.

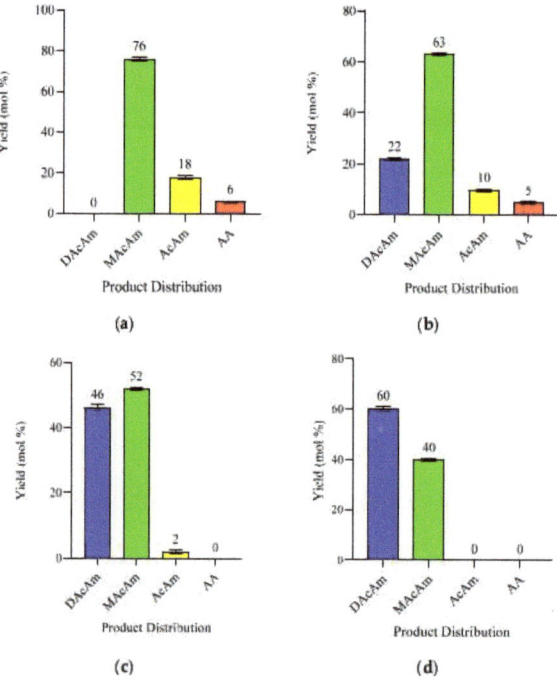

Figure 8. Products distributions of dehalogenation of dichloroacetamide performed in different reaction media. [DAcAm] = 8.8 mM, [NaBH$_4$] = 0.176 M, 0.10 g catalyst, reaction medium: (**a**) deaerated water; (**b**) 8:2 (EtOH:H$_2$O); (**c**) 8:2 (CH$_3$CN:H$_2$O); (**d**) 8:2 (2-Propanol:H$_2$O).

The results in Table 4 suggest that:

I. The BH_4^- will first reduce the surface hydroxides/oxides of the ZVI-NPs. Then, it is expected to form $(ZVI\text{-}NP)\text{-}H_{n+m}^{(n-m)-}$ via reactions (2) and (3) [31]. In the absence of an oxidizing agent, these reactions are followed by reactions (4) and/or (5) [31]. In the presence of an oxidizing substrate, reactions (4) and (5) always compete with the reduction process.

$$ZVI\text{-}NP + (n/4)BH_4^- + (3/4)nH_2O \rightarrow (ZVI\text{-}NP)\text{-}H_n^{n-} + (n/4)B(OH)_3 + (3n/4)H^+ \quad (2)$$

$$(ZVI\text{-}NP)\text{-}H_n^{n-} + mH_2O \rightleftarrows (ZVI\text{-}NP)\text{-}H_{n+m}^{(n-m)-} + mOH^- \quad (3)$$

$$(ZVI\text{-}NP)\text{-}H_{n+m}^{(n-m)-} \rightarrow (ZVI\text{-}NP)\text{-}H_{n+m-2}^{(n-m)-} + H_2 \quad (4)$$

$$(ZVI\text{-}NP)\text{-}H_{n+m}^{(n-m)-} + H_2O \rightarrow (ZVI\text{-}NP)\text{-}H_{n+m-}^{(n-m-1)-} + H_2 + OH^- \quad (5)$$

II. The dehalogenation of the chloro-acetamides by BH_4^- is expected to follow an analogous mechanism to the dechlorination of $Cl_3CCO_2^-$ [31]. As $B(OH)_3$ formed in reaction (2) is a buffer with a pK_b = 4.9 and as OH^- is formed in reactions (3) and (5), the reaction media are always slightly alkaline. The first step in the de-halogenation process is expected to follow either reaction (6), which is a hydrogen transfer process, or reaction (7), which is an electron transfer process; in both, the radical $ClCHCONH_2\cdot$ is formed.

$$\{(ZVI\text{-}NP)\text{-}H_{n+m}\}^{(n-m)-} + Cl_2CHCONH_2 \rightarrow \{(ZVI\text{-}NP)\text{-}H_{n+m-1}\}^{(n-m)-} + Cl^- + H^+ + ClCHCONH_2\cdot \quad (6)$$

$$\{(ZVI\text{-}NP)\text{-}H_{n+m}\}^{(n-m)-} + Cl_2CHCONH_2 \rightarrow \{(ZVI\text{-}NP)\text{-}H_{n+m}\}^{(n-m-1)-} + Cl^- + ClCHCONH_2\cdot \quad (7)$$

This radical is expected to react with the $\{(ZVI\text{-}NP)\text{-}H_{n+m-1}\}^{(n-m)-}/\{(ZVI\text{-}NP)\text{-}H_{n+m}\}^{(n-m-1)-}$ nanoparticles via reactions (8) and/or (9). It should be noted that analogous reactions to reaction (8) [42,43] and (9) [44] were reported to be fast.

$$\{(ZVI\text{-}NP)\text{-}H_{n+m-1}\}^{(n-m)-} + ClCHCONH_2\cdot \rightarrow \{(ZVI\text{-}NP)\text{-}H_{n+m-1}\}\text{-}CHClCONH_2^{(n-m)-} \quad (8)$$

$$\{(ZVI\text{-}NP)\text{-}H_{n+m-1}\}^{(n-m)-} + ClCHCONH_2\cdot \rightarrow \{(ZVI\text{-}NP)\text{-}H_{n+m-2}\}^{(n-m)-} + CH_2ClCONH_2 \quad (9)$$

The $\{(ZVI\text{-}NP)\text{-}H_{n+m-1}\}\text{-}CHClCONH_2^{(n-m)-}$ formed in reaction (8) might decompose via reactions (10)–(13):

$$\{(ZVI\text{-}NP)\text{-}H_{n+m-1}\}\text{-}CHClCONH_2^{(n-m)-} + H_2O \rightarrow \{(ZVI\text{-}NP)\text{-}H_{n+m-1}\}^{(n-m-1)-} + CH_2ClCONH_2 + OH^- \quad (10)$$

$$\{(ZVI\text{-}NP)\text{-}H_{n+m-1}\}\text{-}CHClCONH_2^{(n-m)-} \rightarrow \{(ZVI\text{-}NP)\text{-}H_{n+m-2}\}^{(n-m)-} + CH_2ClCONH_2 \quad (11)$$

$$\{(ZVI\text{-}NP)\text{-}H_{n+m-1}\}\text{-}CHClCONH_2^{(n-m)-} \rightarrow \{(ZVI\text{-}NP)\text{-}H_{n+m-1}\}\text{=}CHCONH_2^{(n-m-1)-} + Cl^- \quad (12)$$

$$\{(ZVI\text{-}NP)\text{-}H_{n+m-1}\}\text{-}CHClCONH_2^{(n-m)-} \rightarrow \{(ZVI\text{-}NP)\text{-}H_{n+m-2}\}\text{=}CHCONH_2^{(n-m)-} + Cl^- + H^+ \quad (13)$$

The $\{(ZVI\text{-}NP)\text{-}H_{n+m-1}\}\text{=}CHCONH_2^{(n-m-1)-}/\{(ZVI\text{-}NP)\text{-}H_{n+m-2}\}\text{=}CHCONH_2^{(n-m)-}$ formed in the latter reactions will decompose via reactions (14)–(17).

$$\{(ZVI\text{-}NP)\text{-}H_{n+m-1}\}\text{=}CHCONH_2^{(n-m-1)-} + 2H_2O \rightarrow (ZVI\text{-}NP)\text{-}H_{n+m-1}\}^{(n-m-3)-} + CH_3CONH_2 + 2OH^- \quad (14)$$

$$\{(ZVI\text{-}NP)\text{-}H_{n+m-2}\}\text{=}CHCONH_2^{(n-m)-} + 2H_2O \rightarrow \{(ZVI\text{-}NP)\text{-}H_{n+m-2}\}^{(n-m-2)-} + CH_3CONH_2 + 2OH^- \quad (15)$$

$$\{(ZVI\text{-}NP)\text{-}H_{n+m-1}\}\text{=}CHCONH_2^{(n-m-1)-} \rightarrow \{(ZVI\text{-}NP)\text{-}H_{n+m-3}\}^{(n-m-1)-} + CH_3CONH_2 \quad (16)$$

$$\{(ZVI\text{-}NP)\text{-}H_{n+m-2}\}\text{=}CHCONH_2^{(n-m)-} \rightarrow \{(ZVI\text{-}NP)\text{-}H_{n+m-4}\}^{(n-m)-} + CH_3CONH_2 \quad (17)$$

III. The observation that in the aqueous and ethanolic media, acetic acid is formed only from the DAcAm and not from the MAcAm proves that partially the

{(ZVI-NP)-H_{n+m-1}}-CHClCONH$_2^{(n-m)-}$ does not decompose via reactions (9)–(11) and that the decomposition of the {(ZVI-NP)-H_{n+m-1}}-CHClCONH$_2^{(n-m)-}$ and/or the {(ZVI-NP)-H_{n+m-1}}=CHCONH$_2^{(n-m-1)-}$/{(ZVI-NP)-H_{n+m-2}}=CHCONH$_2^{(n-m)-}$ intermediates, or at least one of them, involves the hydrolysis of the amide group prior to the loss of the second chloride or prior to the decomposition of the ZVI=CH- bond (reactions (17) or (18)). The latter bond type is not formed during the dechlorination of MAcAm.

$$\{(ZVI\text{-}NP)\text{-}H_{n+m-1}\}\text{-}CHClCONH_2^{(n-m)-} + H_2O \rightarrow \{(ZVI\text{-}NP)\text{-}H_{n+m-1}\}\text{-}CHClCOOH^{(n-m)-} + NH_3 \quad (18)$$

$$\{(ZVI\text{-}NP)\text{-}H_{n+m-2}\}=CHCONH_2^{(n-m)-} + H_2O \rightarrow \{(ZVI\text{-}NP)\text{-}H_{n+m-2}\}=CHCOOH^{(n-m)-} + NH_3 \quad (19)$$

The MAcAm product formed in the dechlorination of DAcAm occurs via one or more of reactions (9)–(11).

IV. The results indicate that the nature of the solvent, though enough water is always present, affects the mechanism of the dehalogenation considerably. At present, the data available do not enable us to determine at what stage of the process this is crucial.

The relative difficulty in performing MAcAm dechlorination in comparison with the dechlorination of DAcAm is attributed to the relative stabilities of the C-Cl bonds.

It was subsequently decided to check whether increasing the amount of the catalyst or the [BH$_4^-$] will enable full dechlorination.

2.3.2. Catalyst Dosing Dehalogenation Experiments

The results presented in Figure 9 indicate that the increase of the catalyst amount suspended in the reaction solution considerably improves the dechlorination process.

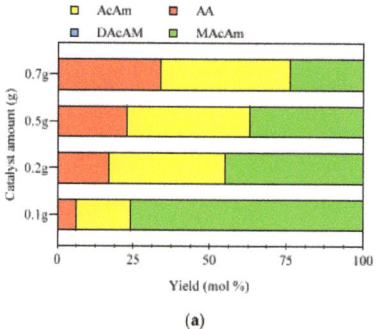

(a)

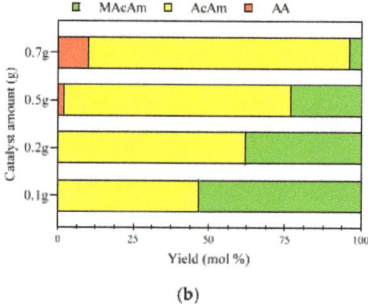

(b)

Figure 9. ZVI@ORMOSIL catalyzed reduction of DAcAm (**a**) and MAcAm (**b**) at increasing catalyst weight. [substrate] = 8.8 mM, [NaBH$_4$] = 0.176 M, reaction medium: water.

The catalyst amount introduced in the reaction has considerable impact on the degree of dechlorination. The ability to reduce the yield of MAcAm to 24% in the reduction of DAcAm, Figure 9a, and the dechlorination by 86% of MAcAm, Figure 9b, is significant. A plausible explanation is that with the decrease in the ratio $[BH_4^-]/[catalyst]$, the charge on the ZVI-NPs decreases, which slows down reactions (4) and (5), which compete with the dechlorination process.

Surprisingly, at the high catalyst concentration, some acetic acid is formed during the dechlorination of MAcAm. This contradicts the discussion above. A plausible explanation is that the intermediates formed during the dechlorination of MAcAm are $\{(ZVI\text{-}NP)\text{-}H_{n+m-k}\}\text{-}(CH_2CONH_2)_k^{(n-m)-}$ and not $\{(ZVI\text{-}NP)\text{-}H_{n+m-1}\}\text{-}CH_2CONH_2^{(n-m)-}$. The number of $-CH_2CO\text{-}NH_2$ bound to a given ZVI-NP affects its properties and a smaller number slows down the ZVI-C bond heterolysis facilitating the amide hydrolysis.

2.3.3. NaBH$_4$ Dosing Dechlorination Experiments

The dependence of the distribution of the products on $[BH_4^-]$ is presented in Figure 10. The results of these experiments indicate that, as expected, an increase in the $[BH_4^-]/[substrate]$ ratio increases the dechlorination yield, though the ratio [dechlorinated products]/$[BH_4^-]$ decreases.

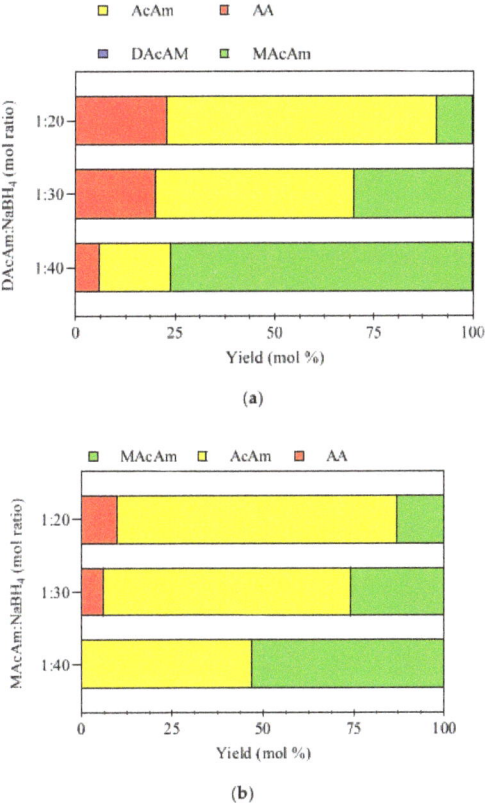

Figure 10. Effect of [NaBH$_4$] on the product distribution. The reaction suspensions in water contained 0.10 g catalyst at room temperature. Reaction time 15 min. (**a**) [DAcAm] = 8.8 mM; (**b**) [MAcAm] = 8.8 mM.

However, here too at high [BH$_4^-$], the dechlorination of MAcAm yields some acetic acid. This is probably due to the increase in the number of H-atoms/hydrides, n+m, bound to the same ZVI-NP coherently. The effect of these hydrogens on the properties of the NPs is similar to that of the -CH$_2$CONH$_2$ groups discussed in the previous section.

In an effort to fully dechlorinate DAcAm and MAcAm, a combination of use of high catalyst amounts, 0.70 g of ZVI@ORMOSIL, and high [BH$_4^-$]/[substrate] = 40:1 was performed. For DAcAm as substrate, the yields were only 7% of MacAm, 57% of AcAm, and 36% of AA. For MAcAm, the yields were 10% of MAcAm, 78% of AcAm, and 12% of AA, i.e., full dechlorination was not achieved. In a last set of experiments, full dechlorination of MAcAm was achieved by performing an identical experiment but delivering the same amount of NaBH$_4$ in two doses. The first dose was allowed to react for 15 min and then the second portion was added for another 15 min. In this experiment, the final products consisted of 63% of acetamide and 37% acetic acid, i.e., 100% dechlorination.

2.3.4. Catalyst Stability Test

To check the recyclability of ZVI@ORMOSIL in the dechlorination of DAcAm, 8 consecutive reactions were performed. For each dehalogenation reaction cycle, the same catalyst was utilized. Before each cycle, the matrix was washed with distilled water and dried. No significant effect on the products distribution was observed (Figure 11).

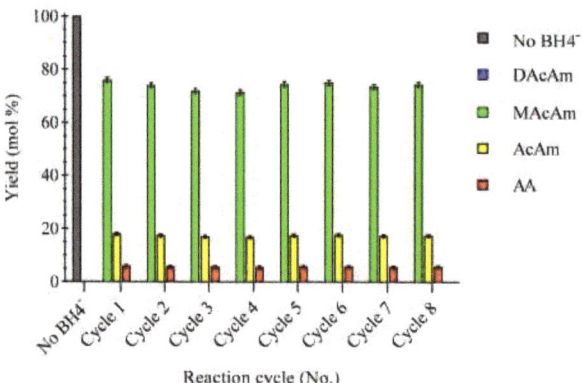

Figure 11. Dechlorination of DAcAm in repeated use of the catalyst. The dechlorination was performed in water following 15 min cycles, [DAcAm] = 8.8 mM, [NaBH$_4$] = 0.176 M, 0.10 g catalyst.

3. Materials and Methods

3.1. Materials

Tetraethylorthosilane (TEOS) > 99%, methyltrimethoxysilane (MTMOS) (97%), zerovalent iron nanoparticles (ZVI-NPs) as iron 25 nm nanopowder > 98.0%, hydrochloric acid (HCl, 37%) and 85% H$_3$PO$_4$ were purchased from Sigma-Aldrich ®(Rehovot, Israel). Sodium borohydride (98%), dichloroacetamide (DAcAm) (>96%), monochloroacetamide (MAcAm) (>96%), acetamide (AcAm) (>97%), acetic acid (AA) (99%) were purchased from Alfa Aesar (Heysham, England). LCMS grade solvents, ethanol (EtOH), acetonitrile (ACN), and 2-propanol, were purchased from Bio-Lab Ltd. (Jerusalem, Israel). All chemicals were used as received. All aqueous solutions were prepared from deionized water purified using a Millipore Milli-Q set up (Merck, Darmstadt, Germany) with a final resistivity of >10 MΩ/cm.

3.2. Synthesis

3.2.1. Preparation of ZVI-NP Suspension

An amount of 0.56 g of 25 nm iron nanoparticles powder was suspended in 10.0 mL of ethanol (AR) resulting in 1.0 M ZVI suspension sealed and stored under nitrogen at 4 °C, equilibrated to room temperature before use.

3.2.2. ZVI@ORMOSIL Synthesis via the Sol-Gel Route

1.0% mol load ZVI@ORMOSIL

The catalyst was prepared by using the two steps acid/base sol-gel synthesis route. Briefly, 37% HCl (62 µL, 2.53 µmol) was dissolved in water (2.72 g, 0.151 mol), and the mixture was added slowly into a premixed solution containing MTMOS (1.556 g, 0.011 mol) and TEOS (5.6 g, 0.027 mol), which were dissolved in ethanol (7.02 g, 0.152 mol). The resulting mixture was homogenized for 15 min. A 2.0% NH_3 solution (1.5 mL, 7.7 mmol) was then added to the mixture dropwise. When the gelation started, 380 µL of 1.0 M ethanolic suspension of ZVI was added and the mixture was stirred vigorously. The wet black gel was kept for 15 days for aging and drying at room temperature. The solid matrix obtained was crushed with a mortar and pestle into a powder and washed with water several times. The washed matrix was then dried and used for the catalytic tests. A tenfold volume of 1.0 M ZVI suspension was used to obtain 10% mol load ZVI@ORMOSIL catalyst.

3.3. Catalyst Characterization

Transmission microscopy images, electron microscopy, and elemental composition were obtained using a 3FE-Tescan ultra high resolution MAIA microscope with an AZTEC microanalysis EDX detector, Oxford Instruments, Colorado, USA. Suspension mean particle size was measured using photon cross correlation spectroscopy (PCCS) with nanophox/R (Sympatec, Germany). The data was analyzed using WINDOX 5. Sample suspensions were sonicated for 10 min in an ultrasonic water bath at room temperature. The measurement was done in a 4 mL plastic cuvette placed in a temperature-controlled water bath. The readings were performed at 632.8 nm. Ethanol (LCMS) RI 1.362, viscosity 1.071 mPas (25 °C).

X-ray diffraction (XRD) measurements were performed with a Bruker (Karlsruhe, Germany) AXS D8 ADVANCE Series II diffractometer equipped with a LynxEye detector (reflection θ–θ geometry, Cu Kα radiation ($\lambda = 0.154$ nm), divergence slit 0.60 mm, anti-scattering slit 8.0 mm). Diffraction data were collected in the angular range of $10° < 2\theta < 80°$, step size 0.05°, and a step time of 0.5 sec/step. N_2 adsorption-desorption isotherms were measured using an APP gold instruments VSorb 2800 model surface area and porosity analyzer, sensitivity 0.010 m^2/g, range 2–500 nm pore size. Specific surface area, average pore size distribution (PSD), pore volume, and adsorption-desorption isotherms were measured in analyses performed using Brunauer–Emmett–Teller (BET) and Barrett–Joyner–Halenda (BJH) methods on a BET surface analyzer. XPS surface analysis was performed using a Thermo Fisher Scientific (East Grinstead, United Kingdom) NEXSA XPS system with monochromatized Al Kα source (400 micron diameter). Pass Energy ("resolution") of 200 eV was used for survey scans to obtain a general surface composition profile and 50 eV for the high-resolution scans that were used for quantitative analysis.

3.4. Catalytic Tests

Dichloroacetamide and monochloroacetamide were taken as model substrates to study the catalytic activity of ZVI@ORMOSIL towards reductive dehalogenation. In a typical catalytic dehalogenation experiment, ZVI@ORMOSIL was weighed (ca. 0.10 g) in a glass vial, 16.90 mg of dichloroacetamide or 12.34 mg of monochloroacetamide (8.8 mM of each substrate) was accurately transferred to the vial. 13 mL of a solvent (deaerated water, ethanol, acetonitrile or 2-propanol) was added, then 0.10 g of $NaBH_4$ was accurately added to the vessel, composing a relative excess ratio of 1:20 (0.132 mmol of substrate/2.64 mmol NaBH$_4$). Finally, 2.0 mL of deuterated water was added to complete a total reaction volume to 15 mL. In a comparative dehalogenation experiment, half of the amount of each substrate was transferred to the same vial in the same preparation method to obtain 1:1 mole/mole mixture, 8.5 mg of DAcAm, and 6.2 mg of MAcAm, yielding 4.4 mM of each substrate in the final reaction solution. The resulting suspensions were stirred for 15 min, and after that, the catalyst was recovered using filtration, and the filtrate was analyzed using HPLC. The reaction samples, standards, and blanks were diluted and filtered before RP-HPLC monitoring for the degradation of haloacetamides. Ammonia amounts are equimolar to the monitored acetic acid for obvious reasons.

HPLC analysis was performed on a Dionex Ultimate 3000 equipped with a Diode Array Detector by Thermo Hypersil-gold C18 150 4.6 mm 3 µm column. Acetic acid, dichloroacetamide, monochloroacetamide, and acetamide were eluted by (0.10% H$_3$PO$_4$; ACN), (4:96) mobile phase, 0.6 mL/min flow, 23 °C column temp, with UV detection at 200 nm, 5.0 µL injection volume. The reaction samples were quenched with a few drops of diluted H$_3$PO$_4$, filtered through 0.22 µm PES or H-PTFE according to the reaction solvent, and all samples were diluted and adjusted to pH 3.0 if necessary, filtered with 0.22 µm PES filter membrane before analysis.

Spectroscopic analysis of ammonia was performed according to a reported procedure [45–48]. Briefly, 0.10 mL of 1.62 M ethanolic solution of phenol was added to 2.5 mL of the filtered reaction solution. After vigorous stirring, 0.10 mL of 0.50% w/v of hexacyanoferrate trihydrate in water was added. After additional stirring, 0.25 mL of the oxidizing solution comprising alkaline 0.64 M citrate trihydrate and 0.12% NaOCl was added, and the standard solutions were prepared likewise. The samples were covered with aluminum foil and kept in the dark at ambient temperature to let the color develop. Absorbances were recorded at 640 nm in 1.0 cm quartz cuvettes. A Varian Cary UV Bio 50 spectrophotometer with dual-beam, Czerny-Turner monochromator, Xenon pulse lamp single source, and dual Si diode detector was used (Agilent technologies, Middleburg, The Netherlands).

4. Conclusions

To conclude, a facile room temperature synthesis of 1% load ZVI@ORMOSIL catalyst via the sol-gel route was performed. The interconnected bottleneck constrictions may contribute to diffusion-controlled rates of dehalogenations. ZVI@ORMOSIL heterogeneous robust catalyst exhibits a good extent of dehalogenation of haloacetamides, producing fully dehalogenated reaction products. Thus, zero valent iron ORMOSIL immobilized nanoparticles prove to be a worthy replacement in reactions catalyzed with rare and expensive metals. The stability of M^0-C formed as an intermediate is a crucial step for determining which solvent mixture may be best for performing the shown reactions.

This report broadens the scope of use of ZVI immobilized in sol-gel as a heterogeneous catalyst for environmental remediation applications. A previously reported use of ZVI as a reductant obtained a noteworthy dehalogenation of haloacetamides, but not to the full dehalogenation extent [25–27]. A complete dehalogenation of DAcAm and MAcAm is reported herein, obtaining fully detoxified acetamide and acetic acid products.

Supplementary Materials: The following are available online at http://www.mdpi.com/2073-4344/10/9/986/s1, Figure S1: (a) Raw wet ORMOSIL gel, (b) crushed ZVI@ORMOSIL gel; Figure S2: Powder diffraction file for Fe0 phases; Figure S3: Powder diffraction file for Fe3O4 phases; Figure S4: (a) pore size distribution for Blank@Ormosil, (b) pore size distribution for 1% ZVI@Ormosil; Table S1: XPS surface analysis elemental composition of 1% ZVI@ORMOSIL.

Author Contributions: Conceptualization, D.M., Y.A., A.B. (Amos Bardea), D.S. and A.B. (Amos Burg); methodology, Y.A., M.M. and D.M.; Funding acquisition, D.M., Y.A. and A.B. (Amos Bardea); Investigation, M.M., Y.A., A.B. (Amos Bardea) and D.M. All authors have read and agreed to the published version of the manuscript.

Funding: This study was enabled in part by a grant from the HIT & Ariel University Joint Research Fund.

Acknowledgments: M. Meistelman thanks the Ariel University for a Ph.D. Fellowship.

Conflicts of Interest: The authors declare no conflict of interest.

References

1. Dragstedt, C.A. The halogenated hydrocarbons: Their toxicity and potential dangers. *AMA Arch. Intern. Med.* **1956**, *97*, 261–262. [CrossRef]
2. Kimura, S.Y.; Vu, T.N.; Komaki, Y.; Plewa, M.J.; Mariñas, B.J. Acetonitrile, and N-Chloroacetamide formation from the reaction of Acetaldehyde and Monochloramine. *Environ. Sci. Technol.* **2015**, *49*, 9954–9963. [CrossRef] [PubMed]
3. Martin, E.T.; McGuire, C.M.; Mubarak, M.S.; Peters, D.G. Electroreductive remediation of halogenated environmental pollutants. *Chem. Rev.* **2016**, *116*, 15198–15234. [CrossRef]
4. Borojovich, E.J.C.; Bar-Ziv, R.; Oster-Golberg, O.; Sebbag, H.; Zinigrad, M.; Meyerstein, D.; Zidki, T. Halo-organic pollutants: The effect of an electrical bias on their decomposition mechanism on porous iron electrodes. *Appl. Catal. B* **2017**, *210*, 255–262. [CrossRef]
5. Letourneau, D.R.; Gill, C.G.; Krogh, E.T. Photosensitized degradation kinetics of trace halogenated contaminants in natural waters using membrane introduction mass spectrometry as an in-situ reaction monitor. *Photochem. Photobiol. Sci.* **2015**, *14*, 2108–2118. [CrossRef]
6. Trojanowicz, M.; Drzewicz, P.; Pańta, P.; Głuszewski, W.; Nałęcz-Jawecki, G.; Sawicki, J.; Sampa, M.H.O.; Oikawa, H.; Borrely, S.I.; Czaplicka, M.; et al. Radiolytic degradation and toxicity changes in g-irradiated solutions of 2,4-dichlorophenol. *Radiat. Phys. Chem.* **2002**, *65*, 357–366. [CrossRef]
7. Kar, P.; Mishra, B.G. Hydrodehalogenation of halogenated organic contaminants from aqueous sources by Pd nanoparticles dispersed in the micropores of pillared clays under transfer hydrogenation condition. *J. Clust. Sci.* **2014**, *25*, 1463–1478. [CrossRef]
8. Heveling, J. Heterogeneous Catalytic chemistry by example of industrial applications. *J. Chem. Educ.* **2012**, *89*, 1530–1536. [CrossRef]
9. Zidki, T.; Bar-Ziv, R.; Green, U.; Cohen, H.; Meisel, D.; Meyerstein, D. The effect of the nano-silica support on the catalytic reduction of water by gold, silver and platinum nanoparticles–nanocomposite reactivity. *Phys. Chem. Chem. Phys.* **2014**, *16*, 15422–15429. [CrossRef] [PubMed]
10. Zidki, T.; Cohen, H.; Meyerstein, D.; Meisel, D. Effect of silica-supported silver nanoparticles on the dihydrogen yields from irradiated aqueous solutions. *J. Phys. Chem. C* **2007**, *111*, 10461–10466. [CrossRef]
11. Adhikary, J.; Meistelman, M.; Burg, A.; Shamir, D.; Meyerstein, D.; Albo, Y. Reductive dehalogenation of monobromo-and tribromoacetic acid by sodium borohydride catalyzed by gold nanoparticles entrapped in sol-gel matrices follows different pathways. *Eur. J. Inorg. Chem.* **2017**, *11*, 1510–1515. [CrossRef]
12. Meistelman, M.; Adhikary, J.; Burg, A.; Shamir, D.; Gershinsky, G.; Meyerstein, D.; Albo, Y. Ag0 and Au0 nanoparticles encapsulated in sol-gel matrices as catalysts in reductive de-halogenation reactions. *Chim. Oggi.* **2017**, *35*, 23–26.
13. Adhikary, J.; Meyerstein, D.; Marks, V.; Meistelman, M.; Gershinsky, G.; Burg, A.; Shamir, D.; Kornweitz, H.; Albo, Y. Sol-gel entrapped Au0-and Ag0-nanoparticles catalyze reductive de- halogenation of halo-organic compounds by BH4–. *Appl. Catal. B-Environ.* **2018**, *239*, 450–462. [CrossRef]
14. Trabelsi, K.; Meistelman, M.; Ciriminna, R.; Albo, Y.; Pagliaro, M. Effective and green removal of trichloroacetic acid from disinfected water. *Materials* **2020**, *13*, 827–828. [CrossRef]
15. Ludwig, J.R.; Schindler, C.S. Catalyst: Sustainable catalysis. *Chemicals* **2017**, *2*, 313–316. [CrossRef]
16. Kaushik, M.; Moores, A. New trends in sustainable nanocatalysis: Emerging use of earth abundant metals. *Curr. Opin. Green Sust.* **2017**, *7*, 39–45. [CrossRef]
17. MacCrehan, W.A.; Bedner, M.; Helz, G.R. Making chlorine greener: Performance of alternative dechlorination agents in wastewater. *Chemosphere* **2005**, *60*, 381–388. [CrossRef]
18. Ibrahem, A.K.; Moghny, T.A.; Mustafa, Y.M.; Maysour, N.E.; El-Din El-Dars, F.M.S.; Hassan, R.F. Degradation of trichloroethylene contaminated soil by zero-valent iron nanoparticles. *ISRN Soil Sci.* **2012**, 1–10. [CrossRef]
19. Bagbi, Y.; Sarswat, A.; Tiwari, S.; Mohan, D.; Pandey, A.; Solanki, P.R. Nanoscale zero-valent iron for aqueous lead removal. *Adv. Mat. Proc.* **2017**, *2*, 235–241. [CrossRef]
20. Fu, F.; Dionysiou, D.D.; Liu, H. The use of zero-valent iron for groundwater remediation and wastewater treatment: A review. *J. Hazard. Mater.* **2014**, *267*, 194–205. [CrossRef]
21. Kharisov, B.I.; Rasika Dias, H.V.; Kharissova, O.V.; Jiménez-Pérez, V.M.; Pérez, B.O.; Flores, B.M. Iron-containing nanomaterials: Synthesis, properties, and environmental applications. *RSC Adv.* **2012**, *2*, 9325–9358. [CrossRef]

22. Guler, U.A. Removal of tetracycline from aqueous solutions using nanoscale zero valent iron and functional pumice modified nanoscale zero valent iron. *J. Environ. Eng. Landsc.* **2017**, *25*, 223–233. [CrossRef]
23. Zha, S.; Cheng, Y.; Gao, Y.; Chen, Z.; Megharaj, M.; Naidu, R. Nanoscale zero-valent iron as a catalyst for heterogeneous Fenton oxidation of amoxicillin. *Chem. Eng. J.* **2014**, *255*, 141–148. [CrossRef]
24. Shea, P.J.; Machacek, T.A.; Comfort, S.D. Accelerated remediation of pesticide-contaminated soil with zerovalent iron. *Environ. Pollut.* **2004**, *132*, 183–188. [CrossRef] [PubMed]
25. Chu, W.; Li, X.; Bond, T.; Gao, N.; Bin, X.; Wang, Q.; Ding, S. Copper increases reductive dehalogenation of haloacetamides by zero-valent iron in drinking water: Reduction efficiency and integrated toxicity risk. *Water Res.* **2016**, *107*, 141–150. [CrossRef] [PubMed]
26. Chen, S.; Chu, W.; Wei, H.; Zhao, H.; Xu, B.; Gao, N.; Yin, D. Reductive dechlorination of haloacetamides in drinking water by Cu/Fe bimetal. *Sep. Purif. Technol.* **2018**, *203*, 226–232. [CrossRef]
27. Chen, S.; Wang, F.; Chu, W.; Li, X.; Wei, H.; Gao, N. Weak magnetic field accelerates chloroacetamide removal by zero-valent iron in drinking water. *Chem. Eng. J.* **2019**, *358*, 40–47. [CrossRef]
28. Bouazizi, N.; Vieillard, J.; Bargougui, R.; Couvrat, N.; Thoumire, O.; Morin, S.; Ladam, G.; Mofaddel, N.; Brun, N.; Azzouz, A.; et al. Entrapment and stability of iron nanoparticles within APTES modified graphene oxide sheets with improved catalytic activity. *J. Alloy. Compd.* **2019**, *771*, 1090–1102. [CrossRef]
29. Sravanthi, K.; Ayodhya, D.; Swamy, P.Y. Green synthesis, characterization, and catalytic activity of 4-nitrophenol reduction and formation of benzimidazoles using bentonite supported zero valent iron nanoparticles. *Mater. Sci. Energy Technol.* **2019**, *2*, 298–307. [CrossRef]
30. Bae, S.; Gim, S.; Kim, H.; Hanna, K. Effect of NaBH4 on properties of nanoscale zero-valent iron and its catalytic activity for reduction of p-nitrophenol. *Appl. Catal. B-Environ.* **2016**, *182*, 541–549. [CrossRef]
31. Meyerstein, D.; Adhikary, J.; Burg, A.; Shamir, D.; Albo, A. Zero-valent iron nanoparticles entrapped in SiO2 sol-gel matrices: A catalyst for the reduction of several pollutants. *Catal. Commun.* **2020**, *133*, 1–5. [CrossRef]
32. Plewa, M.J.; Muellner, M.G.; Richardson, S.D.; Fasano, F.; Buettner, K.M.; Woo, Y.; McKague, A.B.; Wagner, E.D. Occurrence, synthesis, and mammalian cell cytotoxicity and genotoxicity of haloacetamides: An emerging class of nitrogenous drinking water disinfection byproducts. *Environ. Sci. Technol.* **2008**, *42*, 955–961. [CrossRef] [PubMed]
33. Chu, W.; Gao, N.; Yin, D.; Krasner, S.W. Formation, and speciation of nine haloacetamides, an emerging class of nitrogenous DBPs, during chlorination or chloramination. *J. Hazard. Mater.* **2013**, *260*, 806–812. [CrossRef] [PubMed]
34. Chu, W.; Li, D.; Gao, N.; Templeton, M.R.; Tan, C.; Gao, Y. The control of emerging haloacetamide DBP precursors with UV/persulfate treatment. *Water Res.* **2015**, *72*, 340–348. [CrossRef] [PubMed]
35. Brinker, C.J.; Scherer, G.W. *Sol-Gel Science: The Physics and Chemistry of Sol-Gel Processing*; Academic Press: Boston, MA, USA, 1990.
36. Andrade, A.L.; Fabris, J.D.; Ardisson, J.D.; Valente, M.A.; Ferreira, J.M.F. Effect of Tetramethylammonium Hydroxide on nucleation, surface modification and growth of magnetic nanoparticles. *J. Nanomater.* **2012**, *15*, 1–10. [CrossRef]
37. Thommes, M.; Kaneko, K.; Neimark, A.V.; Olivier, J.P.; Rodriguez-Reinoso, F.; Rouquerol, J.; Sing, K.S.W. Physisorption of gases, with special reference to the evaluation of surface area and pore size distribution. *Pure Appl. Chem.* **2015**, *87*, 1–19. [CrossRef]
38. Grosman, A.; Ortega, C. Capillary condensation in porous materials hysteresis and interaction mechanism without pore blocking/percolation process. *Langmuir* **2008**, *24*, 3977–3986. [CrossRef]
39. Alothman, Z.A. A review: Fundamental aspects of silicate mesoporous materials. *Materials* **2012**, *5*, 2874–2902. [CrossRef]
40. Santos, A.M.; Vasconcelos, W.L. Obtention of nanostructured silica glass by sol-gel process with incorporation of lead compounds. *Mater. Res.* **1999**, *2*, 201–204. [CrossRef]
41. Lenza, R.F.S.; Vasconcelos, W.L. Preparation of silica by sol-gel method using formamide. *Mater. Res.* **2001**, *4*, 189–194. [CrossRef]
42. Rusonik, I.; Cohen, H.; Meyerstein, D. Cu (I) (2,5,8,11-tetramethyl-2,5,8,11-tetraazadodecane) as a catalyst for Ullmann's reaction. *J. Chem. Soc. Dalton Trans.* **2003**, *10*, 2024–2028. [CrossRef]
43. Zidki, T.; Cohen, H.; Meyerstein, D. Reactions of alkyl-radicals with gold and silver nanoparticles in aqueous solutions. *Phys. Chem. Chem. Phys.* **2006**, *8*, 3552–3556. [CrossRef] [PubMed]

44. Bar-Ziv, R.; Zidki, T.; Zilbermann, I.; Yardeni, G.; Meyerstein, D. Effect of hydrogen pretreatment of platinum nanoparticles on their catalytic properties: Reactions with alkyl radicals–a mechanistic study. *Chem. Cat. Chem.* **2016**, *8*, 2761–2764. [CrossRef]
45. Solorzano, L. Determination of ammonia in natural waters by the phenol hypochlorite. *Method. Limnol. Oceanogr.* **1969**, *14*, 799–801. [CrossRef]
46. Park, G.; Oh, H.; Ahn, S. Improvement of the ammonia analysis by the phenate method in water and wastewater. *B Korean Chem. Soc.* **2009**, *30*, 2032–2038. [CrossRef]
47. Rhine, E.D.; Sims, G.K.; Mulvaney, R.L.; Pratt, E.J. Improving the Berthelot reaction for determining ammonium in soil extracts and water. *Soil Sci. Soc. Am. J.* **1998**, *62*, 473–480. [CrossRef]
48. Kimble, K.W.; Walker, J.P.; Finegold, D.N.; Asher, S.A. Progress toward the development of a point-of-care photonic crystal ammonia sensor. *Anal. Bioanal. Chem.* **2006**, *385*, 678–685. [CrossRef]

© 2020 by the authors. Licensee MDPI, Basel, Switzerland. This article is an open access article distributed under the terms and conditions of the Creative Commons Attribution (CC BY) license (http://creativecommons.org/licenses/by/4.0/).

Article

Effect of the Gallium and Vanadium on the Dibenzothiophene Hydrodesulfurization and Naphthalene Hydrogenation Activities Using Sulfided NiMo-V$_2$O$_5$/Al$_2$O$_3$-Ga$_2$O$_3$

Esneyder Puello-Polo [1], Yina Pájaro [2] and Edgar Márquez [3],*

[1] Grupo de Investigación en Oxi/Hidrotratamiento Catalítico y Nuevos Materiales,
 Programa de Química-Ciencias Básicas Universidad del Atlántico, Puerto Colombia 081001, Colombia;
 esneyderpuello@mail.uniatlantico.edu.co
[2] Grupo de Investigación en Farmacia Asistencial y Farmacología (GIFAF), Facultad de Química y Farmacia
 Universidad del Atlántico, Puerto Colombia 081001, Colombia; yinapagon@yahoo.es
[3] Grupo de Investigación en Química y Biología, Departamento de Química y Biología, Universidad del Norte,
 Barranquilla 081007, Colombia
* Correspondence: ebrazon@uninorte.edu.co

Received: 4 July 2020; Accepted: 27 July 2020; Published: 7 August 2020

Abstract: The effect of Ga and V as support-modifier and promoter of NiMoV/Al2O3-Ga2O3 catalyst on hydrogenation (HYD) and hydrodesulfurization (HDS) activities was studied. The catalysts were characterized by elemental analysis, textural properties, XRD, XPS, EDS elemental mapping and High-resolution transmission electron microscopy (HRTEM). The chemical analyses by X-ray Fluorescence (XRF) and CHNS-O elemental analysis showed results for all compounds in agreement, within experimental accuracy, according to stoichiometric values proposed to Mo/Ni = 6 and (V+Ni)/(V+Ni+Mo) = 0.35. The sol-gel synthesis method increased the surface area by incorporation of Ga^{3+} ions into the Al$_2$O$_3$ forming Ga-O-Al bonding; whereas the impregnation synthesis method leads to decrease by blocking of alumina pores, as follows NiMoV/Al-Ga(1%-I) < NiMoV/Al-Ga(1%-SG) < NiMo/Al$_2$O$_3$ < Al$_2$O$_3$-Ga$_2$O$_3$(1%-I) < Al$_2$O$_3$-Ga$_2$O$_3$(1%-SG) < Al$_2$O$_3$, propitiating Dp-BJH between 6.18 and 7.89 nm. XRD confirmed a bulk structure typical of (NH$_4$)$_4$[NiMo$_6$O$_{24}$H$_6$]•5H$_2$O and XPS the presence at the surface of Mo^{4+}, Mo^{6+}, Ni$_x$S$_y$, Ni^{2+}, Ga^{3+} and V^{5+} species, respectively. The EDS elemental mapping confirmed that Ni, Mo, Al, Ga, V and S are well-distributed on Al$_2$O$_3$-Ga$_2$O$_3$(1%-SG) support. The HRTEM analysis shows that the length and stacking distribution of MoS$_2$ crystallites varied from 5.07 to 5.94 nm and 2.74 to 3.58 with synthesis method (SG to I). The results of the characterization sulfided catalysts showed that the synthesis method via impregnation induced largest presence of gallium on the surface influencing the dispersion V^{5+} species, this effect improves the dispersion of the MoS$_2$ phase and increasing the number of active sites, which correlates well with the dibenzothiophene HDS and naphthalene HYD activities. The dibenzothiophene HDS activities with overall pseudo-first-order rate constants' values (k$_{HDS}$) from 1.65 to 7.07 L/(h·mol·m^2) follow the order: NiMoV-S/Al-Ga(1%-I) < NiMo-S/Al$_2$O$_3$ < NiMoV-S/Al-Ga(1%-SG), whereas the rate constants' values (k) of naphthalene HYD from 0.022 to 2.23 L/(h·mol·m^2) as follow: NiMoV-S/Al-Ga(1%-SG) < NiMo-S/Al$_2$O$_3$ < NiMoV-S/Al-Ga(1%-I). We consider that Ga and V act as structural promoters in the NiMo catalysts supported on Al$_2$O$_3$ that allows the largest generation of BRIM sites for HYD and CUS sites for DDS.

Keywords: gallium; vanadium; hydrodesulfurization; hydrogenation; synthesis method

1. Introduction

The hydrotreating processes (HDT) uses hydrogenolysis and hydrogenation reactions to remove contaminants such as sulfur, nitrogen, oxygen and metals, and saturate hydrocarbons from liquid petroleum fractions within an oil refinery [1,2]. Catalytic hydrotreating depends largely on the origin of the feed, the operating conditions and the nature of the catalyst with the purpose of increase the quality of transportation fuels [3]. The current generation of hydrotreating catalysts are alumina supported Ni(Co) promoted Mo(W) sulfides; however, these catalysts have some defects, such as the difficulty in its sulfurization and the strong interaction between support and active species [4]. In these catalysts, the models by Topsøe and Chianelli propose that the active sites in these reactions are attributed to sites located on the edges, corners, BRIM or RIM [5,6]. The new regulations nowadays which aim at a severe oil feedstock specifications represent a challenging task for oil refineries [7]. In Colombia, the Ministries of Mines and Energy, in 2014 issued regulations that lead to improved quality of diesel in terms of sulfur and polycyclic aromatic content[8]. To achieve these regulations, in the last decades, efforts have been tried to improve the catalytic properties of traditional catalysts, such as changing the active phase and promoter, varying the preparation method and modifying the support [9,10]. In this sense, many studies describe the influence of support on the performance of HDT catalysts, because their interaction with the active phase determines the morphology, dispersion, sulfur lability, mobility and stability of the corresponding metal site [11]. Usually, the alumina is the catalytic support most used HDT, since it has excellent mechanical, low cost and ability to provide dispersion properties [12]. However, the active components are loaded on it through an impregnation method, which leads to a calcination step that causes the formation of Ni(Co)Mo(W)-AlO$_4$ species (compounds not active in HDT). Hence, recent studies have shown use alumina-modifier elements such as boron, fluorine, phosphorus, silicon, zeolites [13,14], magnesium [15], titanium [16], zinc [17], which could increase the dispersion of the active phase and decrease the active phase(promoter)–support interactions. By taking into account of these limitations, recent studies have shown as potential active phase for such applications the Anderson type polyoxomolybdates [18]. The planar structure of the Anderson type polyoxomolybdates is a relevant factor, producing an active surface with an ordered distribution and uniform deposition, besides the suppression of calcination steps during the activation process could avoid the decrease of active Ni(Co) which favors the synergic effect, doing it an interesting alternative to HDS traditional systems [19].

The addition of gallium and vanadium for the preparation NiMo/Al catalysts has been reported by De los Reyes, who suggested that the Ga^{3+} added to alumina increased HDS activity [20,21]. Cimino and Lo Jacono reported the modification of Ni tetrahedral/octahedral ratio in Ni/Al$_2$O$_3$ catalysts by the addition of Ga^{3+} to alumina [22]. Zepeda et al. demonstrated that the addition of Ga has a strong effect on the CUS, improving the HYD mechanism in the HDS reaction [23]. Petre et al. found that the acidity of the support was modified with Ga, increasing its HDS activity and modifying the desulfurization (DDS)/HYD selectivity [24]. Altamirano et al. reported that the addition of Ga inhibit the formation of Ni(Co)Al$_2$O$_4$ and it improves the sulfiding of the active species [20].

On the other hand, studies have shown that small amounts of vanadium to hydrotreating catalyst leads to the increase in the support acidity as reported by De Jonghe et al. in his study on toluene HYD using V-NiMo catalysts [25]. In this sense, Rankell and Rollman showed that VSx was active for HDT [26]. Lacroix [27] prepared VSx and it was more active in HDS than that MoS$_2$ and WS$_2$, respectively. Paulino et al. have obtained V-based catalysts promoting the HDS and HDN of LCO (7 times greater activities than MoS$_2$) [28,29]. Escalante et al. showed that Al-MCM-41-supported V sulfides catalysts presented the highest formation of hydrogenated products in the thiophene HDS due to the support nature with lower Si/Al(Zr) ratios [30].

In this regard, seeing the importance of Ga and V separately to hydrotreating reactions the present work reports the effect of the addition of Ga and V in NiMoV/Al$_2$O$_3$-Ga$_2$O$_3$ catalyst by two different preparation methods looking to enhance their desulfurization and hydrogenating properties on the HDS and HYD activity of dibenzothiophene and naphthalene.

2. Results and Discussion

2.1. Chemical Analysis

Table 1 shows the experimental chemical analyses of Al_2O_3-Ga_2O_3(1%-SG), Al_2O_3-Ga_2O_3(1%-I), NiMo/Al_2O_3, NiMoV/Al-Ga(1%-SG) and NiMoV/Al-Ga(1%-I) by XRF. The relative deviations between experimental and theoretical could relate to the synthesis procedures. Nevertheless, the experimental accuracy of Mo/Ni and (V+Ni)/(V+Ni+Mo) are in agree with the composition nominal proposed, i.e., 6 and 0.35, respectively [31].

Table 1. Experimental Composition (wt%) and textural properties of supports and NiMoV/Al-Ga(1%-x) catalyst varying the synthesis method (x= SG and I).

Solid	Experimental Composition (wt%)-XRF						Textural Properties				
	Mo	Ni	Ga_2O_3	V_2O_5	$\frac{V+Ni}{Mo+V+Ni}$	$\frac{Mo}{Ni}$	S_{BET}	S_{ext} (m^2/g)	S_{micro}	V_p (cm^3/g)	D_p (nm)
Al_2O_3							265	258	7	0.57	7.68
Al_2O_3-Ga_2O_3(1%-SG)			1.46				259			0.52	6.97
Al_2O_3-Ga_2O_3(1%-I)			1.51				238			0.45	6.78
NiMo/Al_2O_3	25.02	2.38				6.5	233	184	49	0.26	7.20
NiMoV/Al-Ga(1%-SG)	24.02	2.38	0.99	6.43	0.31	6.2	172	160	12	0.37	7.89
NiMoV/Al-Ga(1%-I)	25.60	2.47	0.94	6.57	0.30	6.3	139	138	1	0.23	6.18

SG: sol-gel synthesis; I: impregnation synthesis; S_{BET}: BET surface area; S_{micro}: micropores surface area; S_{ext}: external surface area; Composition Nominal: 20 wt% Mo, 2.04 wt%, 5.5 wt% V_2O_5, 1 wt% Ga_2O_3; (V + Ni)/(Mo + V + Ni) = 0.35, Ni/Mo = 6.

2.2. Textural Properties

All the N_2 physisorption isotherms shown in Figure 1 are type IV in the IUPAC classification [32]. The hysteresis loops showed that Al_2O_3 is type H1 due to uniform mesopores and, Al_2O_3-Ga_2O_3(1%-SG), Al_2O_3-Ga_2O_3(1%-I) and NiMoV/Al-Ga(1%-SG) types a combination of H1 and H2 related to the ink-bottle and uniform type of mesopores, respectively; while NiMoV/Al-Ga(1%-I) displayed a combination of H2 and H3 due to ink-bottle and laminar type mesopores. The isotherms behavior demonstrates the influence of the synthesis method on textural properties of the catalysts [32].

The Table 1 shows that the textural characteristics of the supports and catalysts using supports obtained by the sol-gel synthesis method are greater than those obtained by the conventional impregnation method due to the migration of the metallic phase (Ni, Mo, Ga and V) into the support pores during the impregnation process and/or synthesis of the material that decreases their pore volume and therefore the surface area [33]. Hence, the overall surface area was found to increase as follows: NiMoV/Al-Ga(1%-I) < NiMoV/Al-Ga(1%-SG) < NiMo/Al_2O_3 < Al_2O_3-Ga_2O_3(1%-I) < Al_2O_3-Ga_2O_3(1%-SG) < Al_2O_3; whereas the pore volume showed to increase as follows: NiMoV/Al-Ga(1%-I) < NiMo/Al_2O_3 < NiMoV/Al-Ga(1%-SG) < Al_2O_3-Ga_2O_3(1%-I) < Al_2O_3-Ga_2O_3(1%-SG) < Al_2O_3. The comparison of the supports shows an increase in the surface area and pore volume with the variation of the synthesis method, which means that the Ga^{3+} ions are incorporated into the Al_2O_3 structure (Ga^{3+} ion > Al^{3+} ion) [34,35].

Thus, when comparing the Al_2O_3 with NiMo/Al_2O_3, NiMoV/Al-Ga(1%-SG) and NiMoV/Al-Ga(1%-I) was observed that the S_{micro}/S_{BET} varies between 0.72 and 26.6 %, which can be attributed to the generation of microporosity induced by the migration of the metallic phase, although it decreases with the presence of Ga and V[36]. All the pore size distribution was unimodal (see Table 1 and Figure 1) with pore diameters located in the range of mesoporous (2–50 nm) [32], with values of BJH mesopores mean size between 6.18 and 7.89 nm in the order: NiMoV/Al-Ga(1%-I) < Al_2O_3-Ga_2O_3(1%-I) < Al_2O_3-Ga_2O_3(1%-SG) < NiMo/Al_2O_3 < Al_2O_3 < NiMoV/Al-Ga(1%-SG) [33].

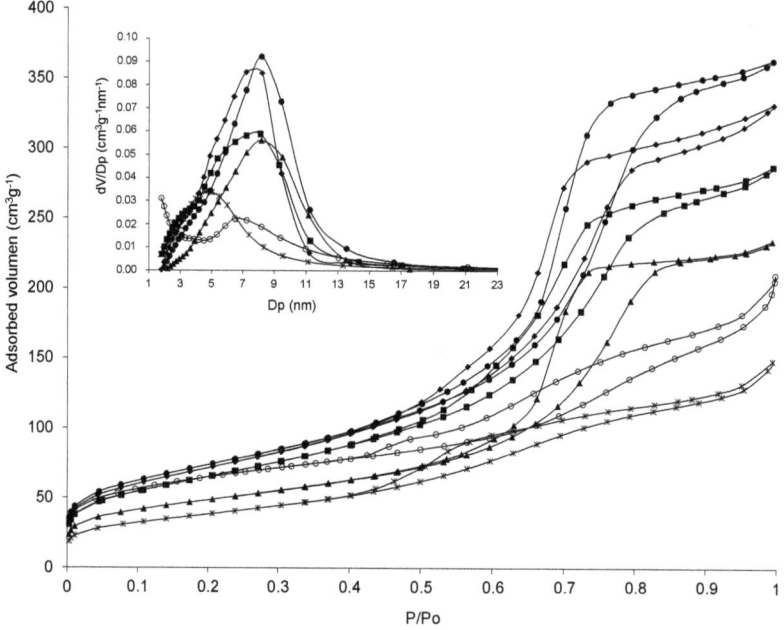

Figure 1. Adsorption/desorption isotherms of N_2 at 77 K and pore size distribution of the supports and NiMoV/Al-Ga(1%-x) catalyst varying the synthesis method (x = SG and I). (•) Al_2O_3; (♦) Al_2O_3-Ga_2O_3(1%-SG); (■) Al_2O_3-Ga_2O_3(1%-I); (○) NiMo/Al_2O_3; (▲) NiMoV/Al-Ga(1%-SG); () NiMoV/Al-Ga(1%-I).

2.3. XRD Analysis

The XRD patterns of oxidic precursors corresponding to Al_2O_3 or Al_2O_3-Ga_2O_3 supported $(NH_4)_4[NiMo_6O_{24}H_6]\bullet 5H_2O$ are shown in Figure 2 [37]. In this figure and regardless of the synthesis method of the support, all the precursors revealed no lines other than those corresponding to $(NH_4)_4[NiMo_6O_{24}H_6]\bullet 5H_2O$ (JCPDS 52-0167) at 2θ = 11.191(100), 15.211(10$\bar{1}$), 16.402(020), 17.548(11$\bar{1}$), 23.772(20$\bar{1}$), 28.587(031), 29.555(211) and γ-Al_2O_3 (JCPDS 10-0425) at 2θ = 67.034(440), 60.899(511), 45.863(400), 39.492(222), 37.604(311), 31.937(220), 19.451(111). The diffraction peaks corresponding to $(NH_4)_4[NiMo_6O_{24}H_6]\bullet 5H_2O$ are narrow, intense and defined, suggesting high crystallinity with greater effect in NiMoV/Al-Ga(1%-SG). Likewise, the XRD pattern of Al_2O_3-Ga_2O_3(1%-SG) and Al_2O_3-Ga_2O_3(1%-SG) shows that the presence of Ga causes a better crystallinity of the Al_2O_3, but a slight shift of diffraction peaks toward larger angles, suggesting a decrease in the interplanar distance which may be related to the changes in porosity in the materials and likewise the precursors. On the other hand, Figure 2, revealed no diffraction lines due to γ-Ga_2O_3 (JCPDS 020-0426, 2θ = 36.191, 64.179) and V_2O_5 (JCPDS 010-0359, 2θ = 20.258, 26.268, 31.138) probably because were well dispersed on the support and/or the crystallites are too small to give XRD signals [31].

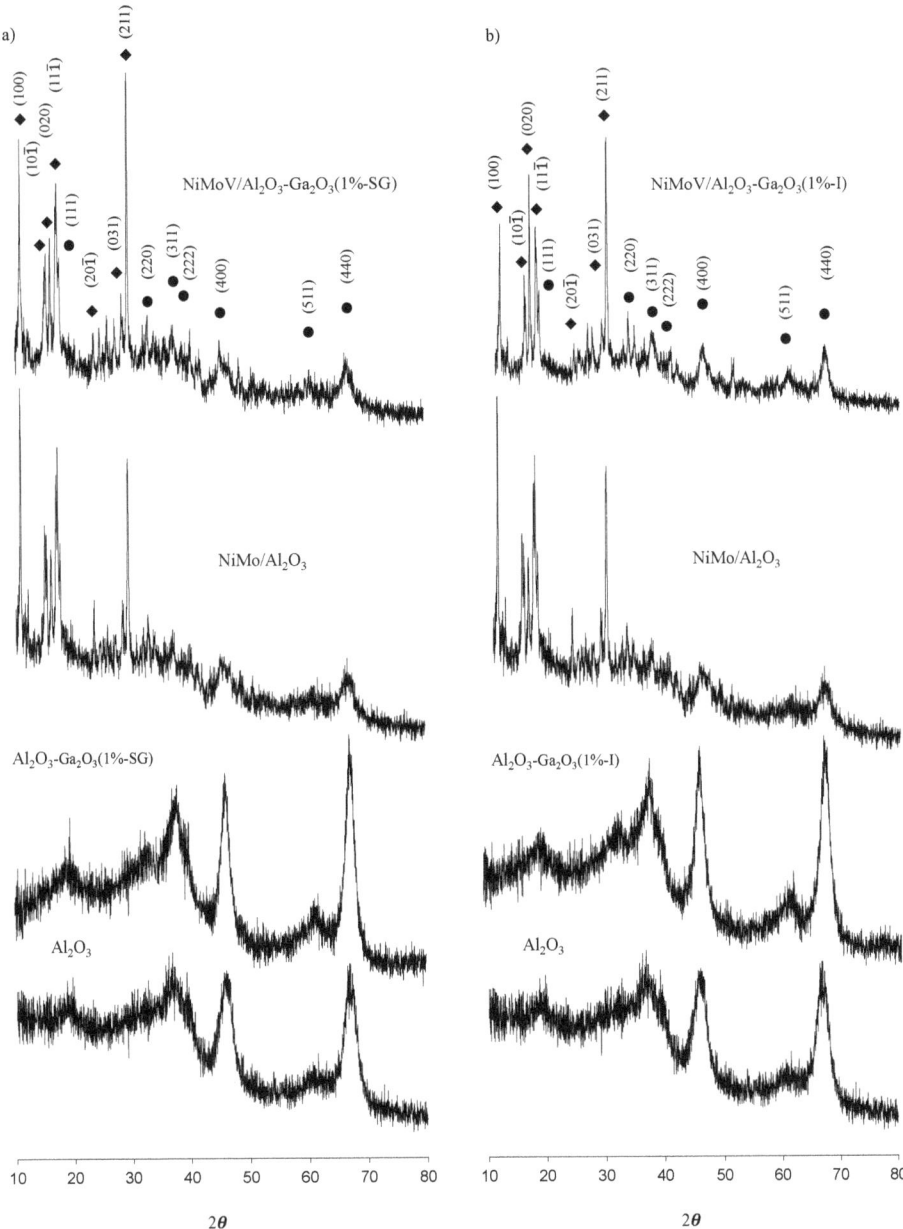

Figure 2. X-ray diffraction patterns of supports and NiMoV/Al-Ga(1%-x) catalyst varying the synthesis method (x = SG and I): (**a**) sol-gel synthesis and (**b**) impregnation synthesis. (●) Al_2O_3; (♦) $(NH_4)_4[NiMo_6O_{24}H_6]\cdot 5H_2O$.

2.4. XPS Analysis

The Figure 2 shows XPS analysis of NiMoV-S/Al-Ga(1%-x). XPS region of Mo $3d_{5/2\text{-}3/2}$ showed signals on the surface of Mo^{4+} (229 eV), Mo^{6+} (232.5 eV) and 2s (226.5 eV), which might be attributable to the Mo sulfide phase, MoO$_3$ and signal of sulfur [38–40]. The higher amount of Mo^{4+} species using the synthesis method of impregnation, suggesting that the increased presence on the surface of gallium and vanadium by impregnation method has a positive effect on reducibility (see Figure 3I). The signal of sulfur in the Mo 3d$_{5/2-3/2}$ region can be confirmed by the presence of three bands in the S 2p$_{3/2}$ region [39]: a signal at 161.7 eV due to terminal disulfide and/or sulfide (S^{2-}), another signal at 163.1 eV corresponding to bridging disulfide (S$_2^{2-}$) ligands and the signal at 168.9 eV, which can be assigned to SO$_4^{2-}$ (see Figure 3III). The signals due to sulfide species are much more important for the catalyst with support modified by impregnation, which showed no presence of sulfates (see Table 2).

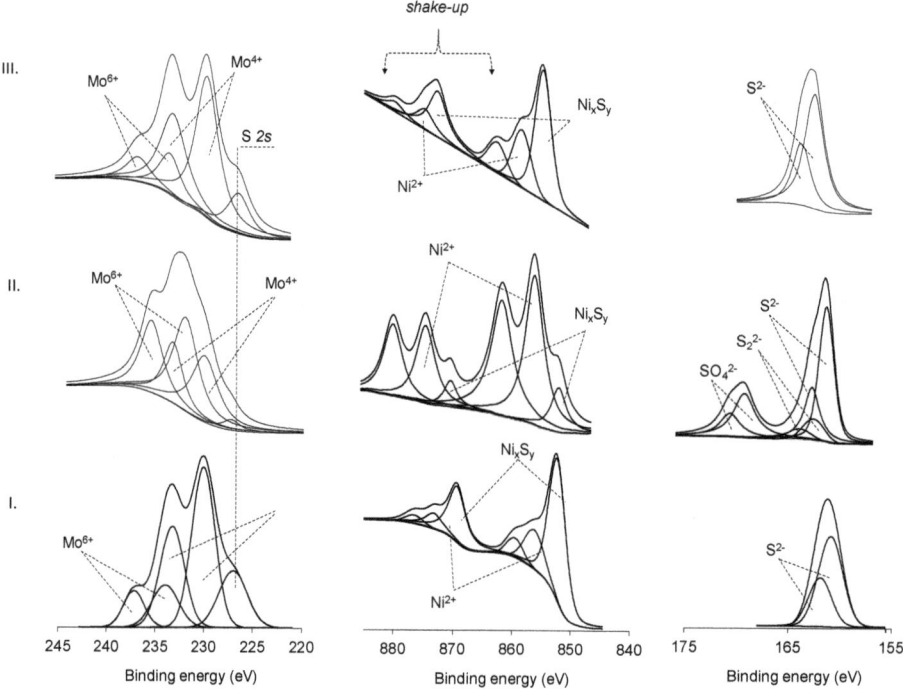

Figure 3. X-ray photoelectron spectra Mo 3d, Ni 2p and S 2p regions of sulfided NiMoV/Al-Ga(1%-x) catalyst varying the synthesis method (x = SG and I): (**I**) NiMo-S/Al$_2$O$_3$; (**II**) NiMoV-S/Al-Ga(1%-SG); (**III**) NiMoV-S/Al-Ga(1%-I).

Meanwhile, the Ni 2p region spectra of Figure 3II shows three Ni 2p$_{3/2-1/2}$ peaks at 853.1–853.6; 856.1–856.7 and 862 eV [41,42]. These signals suggest, respectively, the presence of NixSy sulfide phases (Ni$_2$S$_3$, Ni$_9$S$_8$ or NiS), the NiMoO$_4$ species, and the strong shake-up lines characteristic of Ni^{2+} species in a Ni-Mo-O matrix. Note that the signal of Ni does not show considerable variation in the type of Ni species on surface independently of the support synthesis method, although the proportion of Ni is greater in the impregnation synthesis method (see Table 2). The XPS spectra in the Ga 3d and V 2p$_{3/2-1/2}$ regions (not shown here), the Ga^{3+} (20.5 eV) and V^{5+} (517.1 eV) signals could be attributable to Ga$_2$O$_3$ and V$_2$O$_5$ with amounts from 0.1 to 0.6 on the surface [43,44].

Table 2. Distribution of Mo, Ni, S, Ga and V oxidation states by XPS and morphology of the MoS$_2$ active phase determined by HRTEM in sulfided NiMoV-S/Al-Ga(1%-x) catalyst varying the synthesis method (x = SG and I).

Catalyst	Mo 3d$_{5/2}$-3d$_{3/2}$		Ni 2p$_{3/2}$-2p$_{1/2}$		S 2p$_{3/2}$-2p$_{1/2}$		MoS$_2$ Characteristics		
	Mo^{4+} 229 eV (%)	Mo^{6+} 232.5 eV (%)	Ni$_x$S$_y$ 853.3 eV (%)	NiMoO$_4$ 856.1 eV (%)	S^{2-} 161.7 eV (%)	S$_2$$^{2-}$ 163.1 eV (%) | SO$_4$$^{2-}$ 168.9 eV (%)	L (nm)	N	(fe/fc)$_{Mo}$
NiMo-S/Al$_2$O$_3$	3.30	1.00	0.46	0.14	6.50					
NiMoV-S/Al-Ga(1%-SG)	2.08	2.82	0.083	0.22	1.09	0.22	0.65	5.07	2.74	6.43
NiMoV-S/Al-Ga(1%-I)	2.13	0.57	0.42	0.18	3.60			5.94	3.58	7.79

L (average length) and N (average stacking degree) of MoS$_2$ crystallites; (fe/fc)$_{Mo}$: estimated fraction of Mo atoms on the edge surface of MoS$_2$ particles.

The previous experimental results may be due to the presence of different V^{5+} species on surface, that is, when the support is modified with Ga, small V rich-aggregates on the surface could predominate depending on their dispersion associated to the solubility of the precursors during synthesis [30]. In the synthesis method by impregnation is possible that, in the Al_2O_3-Ga_2O_3(1%-I) support, a large number of Ga^{3+} sites will be available on the surface even after impregnating the V, which is reflected in the good reducing and sulfiding of the Anderson type polyoxomolybdates as seen in the NiMoV-S/Al-Ga(1%-I) catalysts. While in Al_2O_3-Ga_2O_3(1%-SG) there a greater dispersion of the V species.

2.5. SEM Analysis with Energy Dispersive X-ray Spectroscopy and Elemental Mapping

The SEM microscopy of the NiMoV-S/Al-Ga(1%-SG) and NiMoV-S/Al-Ga(1%-I) catalysts showed that the morphologies consist of particle cumulus with irregular geometries, being smaller particles for NiMoV-S/Al-Ga(1%-SG) as displayed the Figure 4. The EDS elemental mapping confirmed the presence of the atoms constituting the catalysts, i.e., Mo, Ni, Al, O, Ga, V and S. These elements are well-distributed on the support as shown by EDS elemental mapping (Figures 4 and 5). However, the Mo, V and S atoms in NiMoV-S/Al-Ga(1%-I) could not disperse in the whole selected area, suggesting that the concentration of these elements is slightly larger in few catalyst zones ("cluster"), hence the differences observed for V can be associated to the solubility of the precursors, suggesting that some V atoms could precipitate on the support surface forming V rich-aggregates [45].

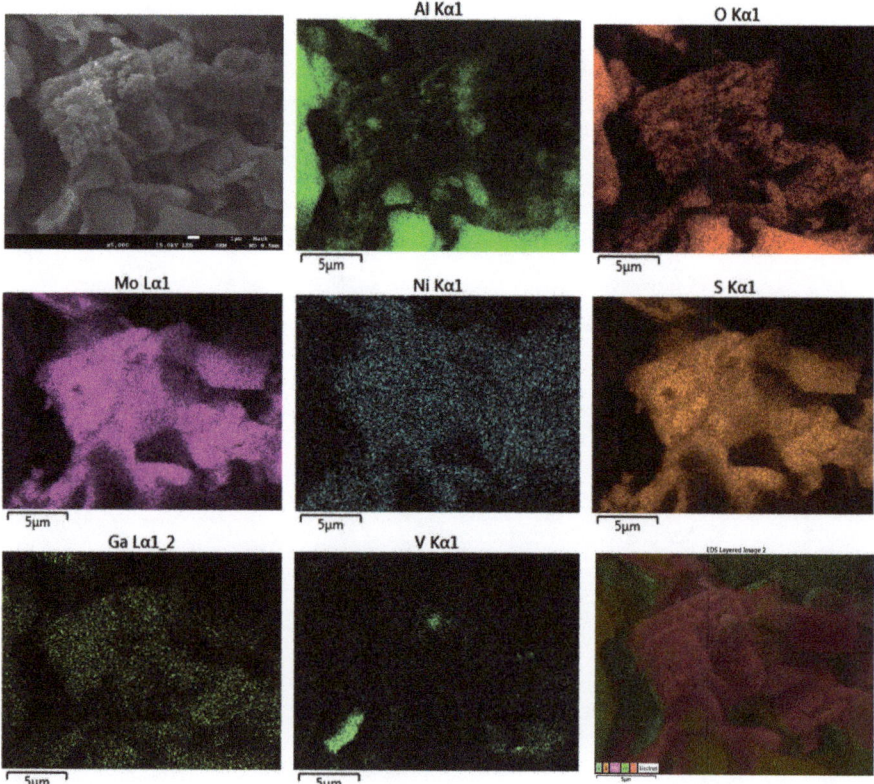

Figure 4. EDS elemental mapping of Mo, Ni, S, Ga, V, Al and O in NiMoV-S/Al-Ga(1%-SG).

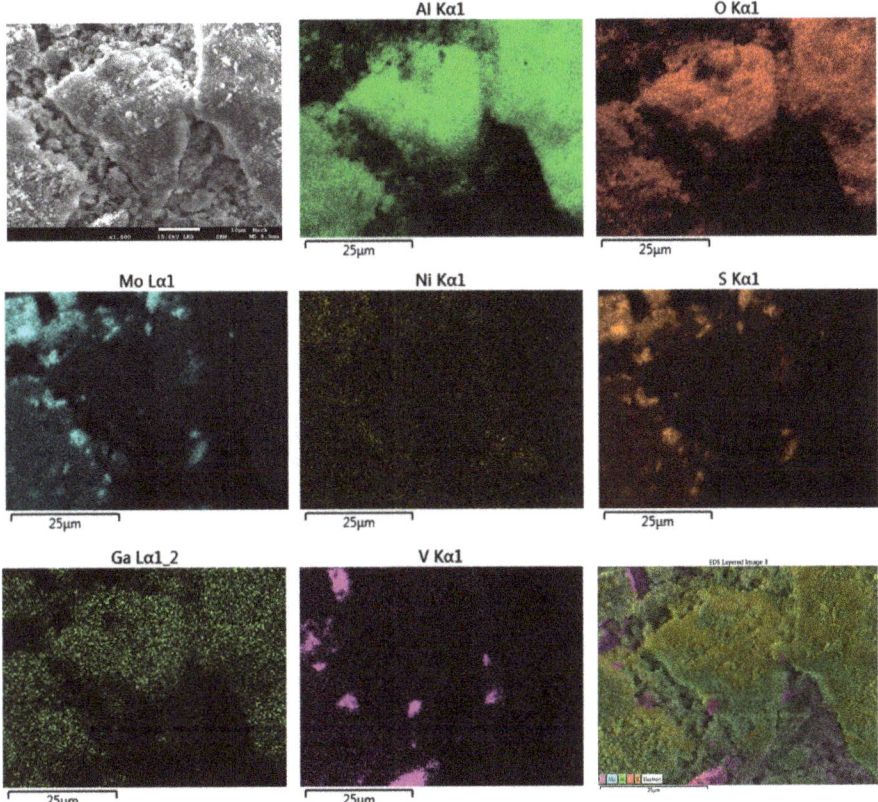

Figure 5. EDS elemental mapping of Mo, Ni, S, Ga, V, Al and O in sulfided NiMoV-S/Al-Ga(1%-I).

2.6. High-Resolution Transmission Electron Microscopy

The HRTEM analysis for NiMoV-S/Al-Ga(1%-SG) and NiMoV-S/Al-Ga(1%-I) is shown in the Figures 6 and 7. The HRTEM micrographs display the presence of homogeneously dispersed MoS_2 crystallites with multi-layers (black thread-like fringes with separation of 0.65 nm characteristic of the basal planes (002)), whose values of D showed higher dispersion of the catalytically active MoS_2 phase in NiMoV-S/Al-Ga(1%-SG) than NiMoV-S/Al-Ga(1%-I) (see Table 2). The Figures 6 and 7 shows that the length and stacking distribution of MoS_2 crystallites changes with the support synthesis method (SG vs. I), which varied from 5.07 to 5.94 nm and 2.74 to 3.58, respectively [46]. Thus, HRTEM image shows that the impregnation method led to the agglomeration of Ni, Mo and V, which was confirmed with the increase in the length and stacking number of the NiMoS phase. This conclusion was also corroborated by EDS elemental mapping (Figure 5). While for the catalyst obtained via sol-gel was improved the dispersion of NiMo species (D = 0.19 vs. 0.20). Meanwhile, the edge-to-corner ratio a MoS_2 slab $(fe/fc)_{Mo}$ increased as the average slab length increased (6.43 to 7.79) [46].

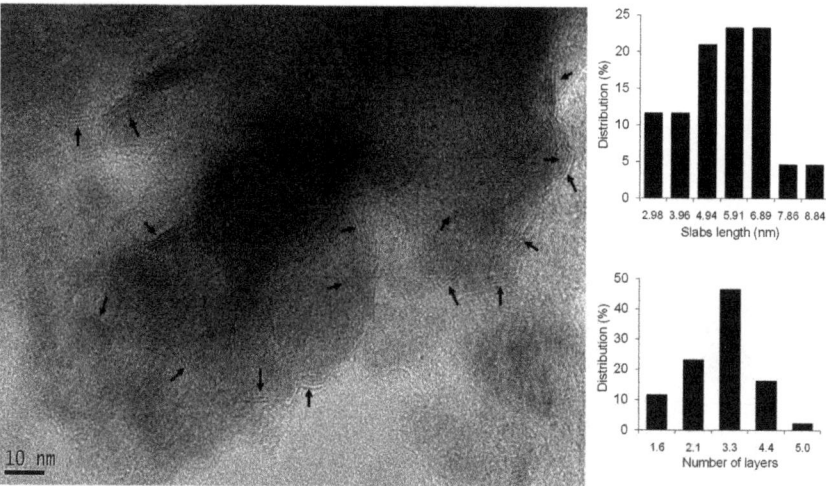

Figure 6. HRTEM micrographs and distributions of slabs length and stacking degree in sulfided NiMoV-S/Al-Ga(1%-SG).

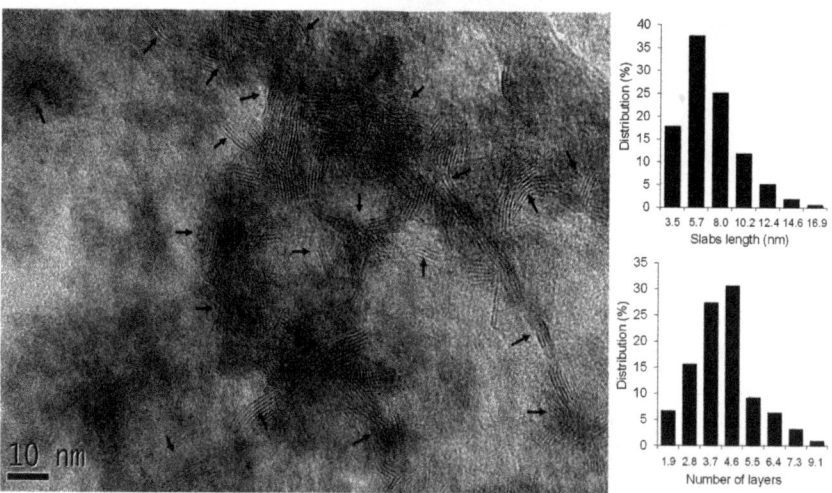

Figure 7. HRTEM micrographs and distributions of slabs length and stacking degree in sulfided NiMoV-S/Al-Ga(1%-I).

2.7. Catalytic Test

The dibenzothiophene HDS activities in function of the product conversion of the NiMo-S/Al_2O_3, NiMoV-S/Al-Ga(1%-SG) and NiMoV-S/Al-Ga(1%-I) catalysts are reported as overall pseudo-first-order rate constants after 6 h of reaction time. Hence, the main reaction products in the HDS of DBT were biphenyl (BP, direct desulfurization route), cyclohexylbenzene (CHB, hydrogenation route) and tetrahydrodibenzothiophene (THDBT), this latter being detected in appreciable amounts at low DBT HDS conversions (<70%) as shows the Figure 8 in the product distributions during reaction times. The catalysts obtained show a large effect of the support synthesis method on intrinsic HDS activities, since the overall activities were found to increase as follows: NiMoV-S/Al-Ga(1%-I) < NiMo-S/Al_2O_3

< NiMoV-S/Al-Ga(1%-SG), i.e., overall pseudo-first-order rate constants' values (k) of 1.65, 2.47 and 7.07 L/(h·mol·m^2), respectively. Likewise, all HDS lead to higher DDS activity (BP formation) with conversions of 85, 87 and 90% for NiMoV-S/Al-Ga(1%-SG), NiMo-S/Al$_2$O$_3$ and NiMoV-S/Al-Ga(1%-I), respectively (see Table 3). The hydrogenation abilities of the catalysts in the HDS reaction was calculated with the HYD/DDS ratios: NiMoV-S/Al-Ga(1%-SG) < NiMo-S/Al$_2$O$_3$ < NiMoV-S/Al-Ga(1%-I). Thus, the introduction of Ga and V to NiMoV-S/Al-Ga(1%-SG) resulted in almost a three-fold increase of the rate constant than their unpromoted analog, while NiMoV-S/Al-Ga(1%-I) was more than half; display that NiMoV-S/Al-Ga(1%-SG) was the most active between the synthesized and tested catalysts. This result suggests that the way to incorporate Ga into the support influences the dispersion of the V^{5+} species (structural promoter) [30]. It is well known that Ga-incorporation into A$_2$O$_3$-framework provides an increased number of acid sites that allow a better dispersion [20]. In the case of NiMoV-S/Al-Ga(1%-I) was observed small V rich-aggregates with greater presence of Ga on the surface that favors good hydrogenating properties NiMoS [20,21].

Table 3. Apparent rate constants of NiMoV-S/Al-Ga(1%-x) catalyst varying the synthesis method (x = SG and I) for DBT hydrodesulfurization (HDS) and naphthalene hydrogenation (HYD) in the reaction network shown in Schemes 1 and 2.

Catalysts	HDS Rate Constants, L/(h·mol·m^2)						HYD Rate Constants, L/(h·mol·m^2)		
	K$_{HDS}$	k$_1$*	k$_2$*	k$_3$* (×10^{-10})	k$_4$*	HYD/DDS	k$_1$*	k$_2$*	k$_3$*
NiMo-S/Al$_2$O$_3$	2.47	2.15	0.314	5.62	46.2	0.15	0.454	0.151	0.164
NiMoV-S/Al-Ga(1%-SG)	7.07	6.34	0.732	0.112	79.1	0.12	0.481	0.022	0.057
NiMoV-S/Al-Ga(1%-I)	1.65	1.40	0.254	0.019	24.5	0.18	1.02	1.61	2.23

kn*, apparent rate constant.

The aforementioned can be confirmed with naphthalene HYD activities, which shows that NiMoV-S/Al-Ga(1%-I) was more active than NiMo-S/Al$_2$O$_3$ and NiMoV-S/Al-Ga(1%-SG) with conversions around 35% as shown in Figure 8. It showed a higher rate of constants' values (k), which varied in a wide range, from 0.022 to 2.23 L/(h·mol·m^2) (see Table 3). Liu et al. reported in their studies of naphthalene HYD on highly-loaded NiMo catalysts that the morphological differences of the agglomerated active components (MoS$_2$ nanoparticles) can be the main reason of their hydrogenation ability and in this sense our NiMoV-S/Al-Ga(1%-I) exhibited much larger MoS$_2$ nanoparticles than NiMoV-S/Al-Ga(1%-SG) as revealed by HRTEM analysis, suggesting that the hydrogenated intermediate tetralin did not need to desorb from the catalyst surface since there might exist enough space for the total hydrogenation reactions; whereas NiMo-S/Al$_2$O$_3$ and NiMoV-S/Al-Ga(1%-SG) showed a higher selectivity towards tetralin, probably due to a function of adsorption nature more than hydrogenation ability of the catalyst [47]. Resuming all the above observations, we consider that Ga and V act as structural promoters in the NiMo catalysts supported on Al$_2$O$_3$ that allows the largest generation of BRIM sites for HYD and coordinative unsaturated sites (CUS) for DDS.

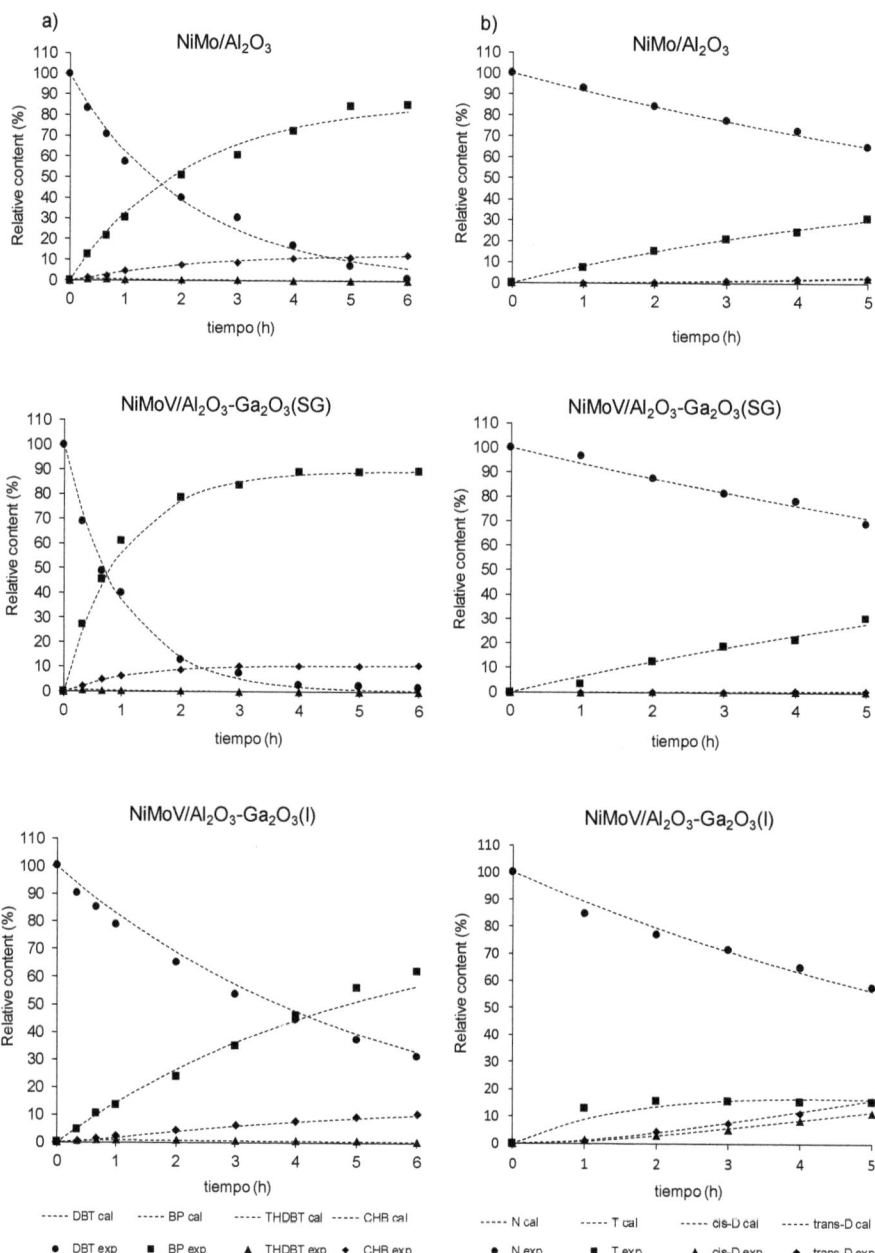

Figure 8. Reaction reactant and products compositions of NiMoV-S/Al-Ga(1%-x) varying the synthesis method (x = SG and I). (**a**) Dibenzothiophene HDS and (**b**) naphthalene HYD. The fitted curves were based on the Equations (1)–(8) of Schemes 1 and 2.

3. Materials and Methods

3.1. Preparation of Alumina Modified with Gallium

The catalytic supports with 1 wt% as Ga_2O_3 was prepared by two different methods that will be identified as Al_2O_3-Ga_2O_3(1%-SG) and Ga_2O_3/Al_2O_3(1%-I). Al_2O_3-Ga_2O_3(1%-SG) was prepared by the one-pot sol-gel synthesis [34]. In a typical experiment, appropriate amounts of aluminum(III)isopropylate ($Al[OCH(CH_3)_2]_3$, 99.8%, Sigma-Aldrich), and gallium(III)acetylacetonate ($Ga[CH_3COCH=C(O-)CH_3]_3$, 99.9%, Sigma-Aldrich) were dispersed in 50 mL of isopropanol (($CH_3)_2CHOH$, 98%, Sigma-Aldrich) under magnetic stirring at about 75–77 °C until obtaining a homogeneous solution. Subsequently, the polymerized solution of $Al[OCH(CH_3)_2]_3$/ $Ga[CH_3COCH=C(O-)CH_3]_3$ was slowly added to a surfactants solution obtained homogenizing 56.5 mmol of tetramethylammonium hydroxide (TMAOH, 25% in H_2O, Sigma-Aldrich) and 18.1 mmol of hexadecyltrimethylammonium bromide (CTMAB, 99 %, Sigma-Aldrich) in 30 mL of deionized water. After that, the pH was adjusted at 8–10 with diluted ammonium hydroxide (NH_4OH, 28.0–30.0% NH_3 basis, Sigma-Aldrich), keeping under stirred for 2 h. The resulting mixture was aged for 48 h without stirring, filtered, washed, dried at 393 K for 12 h, pulverized and calcined at 883 K for 6 h.

In the other hand, The Ga_2O_3/Al_2O_3(1%-I) was prepared by impregnation over pore volume adding dropwise to a flask containing 5 g of Al_2O_3 (alumina was obtained as above-mentioned without adding gallium) an aqueous solution of $Ga[CH_3COCH=C(O-)CH_3]_3$ at 353 K, under stirring and pH 6. after removing the solvent by evaporation, the as-made sample was dried at 383 K for 12 h and then it was calcined at 883 K for 6 h.

3.2. Preparation of Catalyst Precursors Supported on Modified Alumina with Gallium

The vanadium was impregnated on 3 g of Al_2O_3-Ga_2O_3(1%-I) and Al_2O_3-Ga_2O_3(1%-SG) over pore volume adding dropwise an acidified aqueous solution of ammonium metavanadate (5.5 wt% V_2O_5, sigma-Aldrich 98%) at 353 K, under stirring and pH 5–6 until that the solvent is removed by evaporation. The mass obtained was dried at 383 K for 12 h and then it was calcined at 773 K for 4 h. Later, three types of catalytic precursors to 20 wt% Mo were synthesized impregnating in excess of pore volume 3 g of V_2O_5/Al_2O_3-Ga_2O_3(1%-I), V_2O_5/Al_2O_3-Ga_2O_3(1%-SG) and Al_2O_3 with an aqueous solution of Anderson ammonium salt (($NH_4)_4[NiMo_6O_{24}H_6]\bullet 5H_2O$) under stirring at 323 K and pH around 5–6, respectively. The impregnation step lasted until the removal of the solvent by evaporation and the mass obtained was further dried at 378 K for 12 h [31].

The $(NH_4)_4[NiMo_6O_{24}H_6]\bullet 5H_2O$ supported on alumina modified with gallium and vanadium will be identified as NiMoV/Al-Ga(1%-x), where x is the synthesis method of the support (SG: sol-gel and I: impregnation). Likewise, NiMoV/Al-Ga(1%-x) sulfided will be identified as NiMoV-S/Al-Ga(1%-x).

3.3. Catalyst Characterization

The elemental analysis for NiMoV/Al-Ga(1%-x) was determined by XRF using a MagixPro PW–2440 Philips instrument. Sulfur elemental analysis was carried out employing a combustion method employing a Fisons EA 1108 CHNS-O analyzer in solids HDS postreaction. The textural properties were determined utilizing the physisorption technique of N_2 at 77 K using a Micromeritics 3FLEXTM instrument. The surface areas of samples were calculated by the Brunauer–Emmett–Teller multipoint method (BET) and, total pore volume and pore size distribution were determined from the adsorption branch of the isotherm using the Barret–Joyner–Halenda (BJH) model [48]. XRD analysis of the samples was carried out using a BRUKER D8 ADVANCE diffractometer with a Cu Kα radiation source (λ = 1.5418 Å) and Ni filter, within the range $5° \leq 2\theta \leq 90°$, step size of 0.02° and acquisition speed of 0.08°/s. Identification of the different phases was made using the JCPDS library [37]. The surface composition of the sulfided catalysts was determined through of X-ray photoelectron spectroscopy (XPS) with a Thermo Scientific K-Alpha spectrometer, equipped with a dual (non-monochromatic) Mg/Al anode, operated at 400 W and under a vacuum better than 10^{-9} torr. Calibration of the instrument was carried out employing the Au $4f_{7/2}$ line at 83.9 eV. The internal referencing of binding energies was made using the dominating Al 2p band of the support at 74,4 eV [49]. Morphology of the samples was observed by scanning electron microscopy (SEM) using a field emission scanning electron microscope (JEOL, model JSM-7800F, Japan) operated at 1 kV. The mapping images and elemental analysis characterization were acquired simultaneously at 15 kV using the energy-dispersive X-ray spectroscopy (EDS) analyzer coupled to the same JEOL 7800F instrument. The average morphology of the MoS_2 active phase in the catalysts was observed in HRTEM images, which were obtained on a JEOL 2010 microscope with a 1.9 Å point-to-point resolution at 200 kV. From 10 to 15 representative micrographs were obtained for each catalyst in high-resolution mode. Typically, the slabs lengths and stacking of least 350-400 crystallites of MoS_2 were measured for each catalyst along with its dispersion, the average fraction of Mo atoms at the MoS_2 edge surface was calculated as suggesting Li et al. [46].

3.4. Catalytic Test

Before the HDS and HYD catalytic reactions, 1.0 g of NiMoV/Al-Ga(1%-x) or NiMo/Al_2O_3 precursor were sulfiding ex-situ passing through them a total flow rate of 0.33 cm^3 min^{-1} CS_2(2 vol%)/heptane and 70 cm^3 min^{-1} of hydrogen at 623 K for 4 h to attain a reproducible and stable state at the surface [36]. The HDS and HYD test conditions were as follows: The dibenzothiophene/hexadecane (80 mL, 500 ppm of S) and naphthalene/hexadecane (80 mL, 0.12 M) solutions were introduced, respectively, into the autoclave (JP Inglobal) with 300 mg of catalyst and then the reactor was purged for three times by N_2 at ambient temperature and thereafter pressured to 3.1 MPa H_2. The mixture was heated from room temperature to 593 K for 6 h under constant stirring, then the dibenzothiophene and naphthalene consumption and products formed during the course of reaction were followed employing a CG-2014-Shimadzu equipped with a flame ionization detector (FID) and 30 m length BP5 capillary column using standards of dibenzothiophene (DBT, Sigma-Aldrich 99%), biphenyl (BP, Sigma-Aldrich ≥ 99%), cyclohexylbenzene (CHB, Sigma-Aldrich ≥ 97%), tetrahydrodibenzothiophene (THDBT, GC-MS), naphthalene (N, Sigma-Aldrich 99%), tetralin (T, Sigma-Aldrich 99%), trans-decalin (trans-D, Sigma-Aldrich 99%), cis-decalin (cis-D, Sigma-Aldrich 99%). The effluents sampling of the reactor occurring at 0, 30, 60, 120, 180, 240, 300 and 360 min.

In the reaction both the alumina supporting, and the V or Ga-modified alumina showed negligible dibenzothiophene and naphthalene conversion. Absence of mass and heat flow transport effects was verified according to established procedures [50,51]. All experiments reported in this work (synthesis protocols, characterizations and catalytic activity measurements) were carried out at least in triplicate. Good reproducibility was verified, better than 10% in all quantitative measurements.

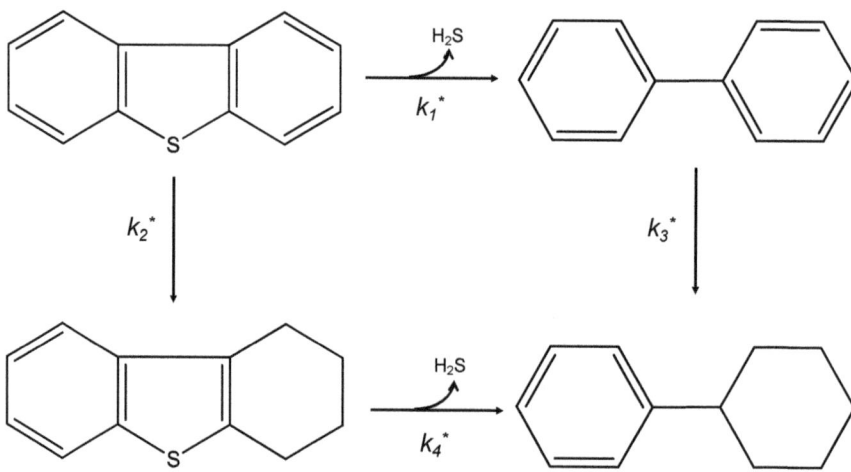

Scheme 1. Semi-empirical kinetic models of HDS [52].

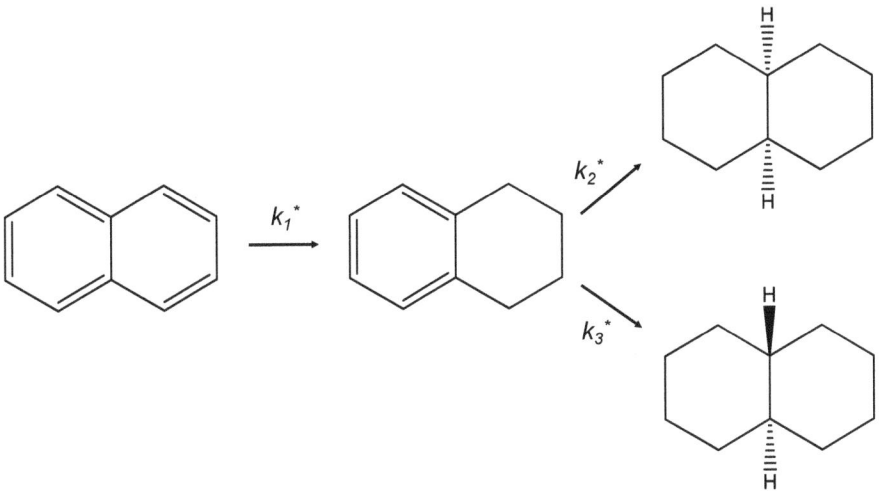

Scheme 2. Semi-empirical kinetic models of HYD [52].

The semi-empirical kinetic models of HDS and HYD were calculated according to the mechanism presented in the Schemes 1 and 2 [52,53], respectively. All the reactions were assumed irreversible due to the excess of hydrogen (considered constant), whereby the reactions are considered of the pseudo-first-order. The HDS kinetic model assumes the existence of active sites for hydrogenation and direct desulfurization, which can be expressed as $R_{Total} = R_{HYD} + R_{DDS}$, namely the overall pseudo-first order rate constant (k) as $k = k_1^* + k_2^*$, where k_1^* and k_2^* are the apparent rate constants for the DDS and HYD routes; while the naphthalene is firstly hydrogenated to tetralin, and later to decalin isomers as final reaction products.

The system of differential equations to calculate the catalytic constants was solved using the Maxima software and the nonlinear parameter estimations were calculated until converged by minimizing the deviation from experimental concentrations using the Levenberg–Marquardt algorithm of the Origin 9 version. In this approach, the apparent rate constants of the dibenzothiophene HDS and naphthalene HYD were calculated as k_1^*, k_2^*, k_3^* and k_4^* (Equations (1)–(4)) and k_1^*, k_2^* and k_3^* (Equations (5)–(8)), which are defined by $k_n^* = k_n K_n$, where k_n and K_n are the intrinsic rate and the equilibrium adsorption constants, respectively.

$$C_{DBT} = C_{DBT_0} \cdot e^{-kt} \tag{1}$$

$$C_{BP} = \frac{C_{DBT_0} \cdot k_{1*}}{k_{3*} - k}\left[e^{-kt} - e^{-k_{3*}t}\right] \tag{2}$$

$$C_{THDBT} = \frac{C_{DBT_0} \cdot k_{2*}}{k_{4*} - k}\left[e^{-kt} - e^{-k_{4*}t}\right] \tag{3}$$

$$C_{CHB} = \frac{C_{DBT_0} \cdot k_{3*} \cdot k_{1*}}{k_{3*} - k}\left[\frac{1}{k_{3*}}e^{-k_{3*}t} - \frac{1}{k}e^{-kt}\right] \\ + \frac{C_{DBT_0} \cdot k_{4*} \cdot k_{2*}}{k_{4*} - k}\left[\frac{1}{k_{4*}}e^{-k_{4*}t} - \frac{1}{k}e^{-kt}\right] + \frac{C_{DBT_0}(k_{1*} + k_{2*})}{k} \tag{4}$$

$$C_N = C_{N_0} \cdot e^{-k_{1*}t} \tag{5}$$

$$C_T = \frac{C_{N_0} \cdot k_{1*}}{(k_{2*} + k_{3*}) - k_{1*}}\left[e^{-k_{1*}t} - e^{-(k_{2*}+k_{3*})t}\right] \tag{6}$$

$$C_{cisD} = C_{N_0} \left[\frac{k_{2*}}{(k_{2*}+k_{3*})} - \frac{k_{1*} \cdot k_{2*}}{(k_{2*}+k_{3*}) \cdot (k_{1*}-k_{2*}-k_{3*})} e^{-(k_{2*}+k_{3*})t} + \frac{k_{2*}}{k_{1*}-k_{2*}-k_{3*}} e^{-k_{1*}t} \right] \quad (7)$$

$$C_{transD} = C_{N_0} \left[\frac{k_{3*}}{(k_{2*}+k_{3*})} - \frac{k_{1*} \cdot k_{3*}}{(k_{2*}+k_{3*}) \cdot (k_{1*}-k_{2*}-k_{3*})} e^{-(k_{2*}+k_{3*})t} + \frac{k_{3*}}{k_{1*}-k_{2*}-k_{3*}} e^{-k_{1*}t} \right] \quad (8)$$

Finally, the HDS and HYD activities of the catalysts are reported as pseudo-first-order rate constant for dibenzothiophene and naphthalene disappearance normalized by the concentration (mol/L) of DBT or N per weight (m_{cat}, g) and surface area (S_{BET}, m^2/g) of the catalyst after ~ 5–6 h of reaction time.

4. Conclusions

In the present work, two NiMoV catalysts supported on Al_2O_3-Ga_2O_3(1%-x) were prepared using two synthesis method different. The changes in the support's composition with gallium and vanadium as promoter resulted in properties different and dispersion of the MoS_2 particles. The evaluation of the NiMoV/Al_2O_3-Ga_2O_3(1%-x) catalysts in the DBT hydrodesulfurization and NP hydrogenation showed that the catalysts presented significant differences in the activity and selectivity of the products. In this respect, the chemical analyses by FRX for the NiMoV/Al_2O_3-Ga_2O_3(x) catalysts showed stoichiometric values Mo/Ni ~ 6 and (V + Ni)/(V + Ni + Mo) = 0.31–0.34, while EDS spectra and elemental mapping confirmed the presence of Ni, Mo, V, Ga, S, Al and O with well-distribution on support. The NiMo/Al_2O_3 and NiMoV/Al_2O_3-Ga_2O_3(1%-SG) showed better textural properties than NiMoV/Al_2O_3-Ga_2O_3(1%-I) with average pore radius between 6.18 and 7.89 nm. XRD confirmed the presence of $(NH_4)_4[NiMo_6O_{24}H_6] \bullet 5H_2O$ and XPS evidenced Mo^{4+}, Mo^{6+}, Ni_xS_y, Ni^{2+}, Ga^{3+} and V^{5+} species, which shown a dependence with the synthesis method of the support, whereas gallium and vanadium oxides were not detected due to the well dispersed on the support, suggesting that the synthesis method via impregnation induced largest presence of gallium on the surface allowing a better dispersion of V^{5+} oxides influencing in the degree of sulfidation. The HRTEM analysis shown that the length and stacking distribution of MoS_2 crystallites changes with the support synthesis method varied from 5.07 to 5.94 nm and 2.74 to 3.58, respectively. The activities as dibenzothiophene HDS overall pseudo-first-order rate constants' values (k_{HDS}) from 1.65 to 7.07 L/(h·mol·m^2), and naphthalene HYD activity with rate constants' values (k) from 0.022 to 2.23 L/(h·mol·m^2). The previous observations allow us to conclude that Ga and V act as structural promoters in the NiMo catalysts supported on Al_2O_3 that contribute in the largest generation of BRIM sites for HYD and CUS sites for DDS.

Author Contributions: Conceptualization, E.P.-P. and E.M.; methodology, E.P.-P.; software, E.P.-P.; validation, E.P.-P., E.M., and Y.P.; formal analysis, Y.P.; investigation, E.P.-P.; resources, Y.P.; data curation, E.P.-P. and E.M.; writing—original draft preparation, E.M.; writing—review and editing, E.P.-P., E.M., and Y.P.; visualization, E.M.; supervision, E.P.-P., E.M., and Y.P.; project administration, E.P.-P. and E.M.; funding acquisition, E.P.-P.; resources, E.P.-P., E.M. and Y.P.. All authors have read and agreed to the published version of the manuscript.

Funding: This research received no external funding.

Acknowledgments: The authors also gratefully the financial support provided by Universidad del Norte, under Project number: 2019-017. Besides, The authors would like to acknowledge financial support to Universidad del Atlántico (through equidad investigativa y "1° convocatoria interna para apoyo al Desarrollo de trabajos de grado en investigación formativa nivel pregrado y postgrado").

Conflicts of Interest: The authors declare no conflict of interests.

References

1. Anderson, J.R.; Boudart, M. *Catalysis: Science and Technology*; Springer: Berlin/Heidelberg, Germany, 1996; ISBN 978-3-642-61040-0.
2. Raṣeev, S.D. *Thermal and Catalytic Processes in Petroleum Refining*; Marcel Dekker: New York, NY, USA, 2003; ISBN 978-0-8247-0952-5.

3. Lødeng, R.; Hannevold, L.; Bergem, H.; Stöcker, M. Catalytic Hydrotreatment of Bio-Oils for High-Quality Fuel Production. In *The Role of Catalysis for the Sustainable Production of Bio-Fuels and Bio-Chemicals*; Triantafyllidis, K.S., Lappas, A.A., Stöcker, M., Eds.; Elsevier: Amsterdam, The Netherlands, 2013; Chapter 11; pp. 351–396. ISBN 978-0-444-56330-9.
4. Debecker, D.P.; Stoyanova, M.; Rodemerck, U.; Gaigneaux, E.M. Preparation of $MoO_3/SiO_2-Al_2O_3$ metathesis catalysts via wet impregnation with different Mo precursors. *J. Mol. Catal. Chem.* **2011**, *340*, 65–76. [CrossRef]
5. Topsøe, H.; Clausen, B.S. Active sites and support effects in hydrodesulfurization catalysts. *Appl. Catal.* **1986**, *25*, 273–293. [CrossRef]
6. Chianelli, R.R.; Siadati, M.H.; De la Rosa, M.P.; Berhault, G.; Wilcoxon, J.P.; Bearden, R.; Abrams, B.L. Catalytic Properties of Single Layers of Transition Metal Sulfide Catalytic Materials. *Catal. Rev.* **2006**, *48*, 1–41. [CrossRef]
7. Babich, I. Science and technology of novel processes for deep desulfurization of oil refinery streams: A review. *Fuel* **2003**, *82*, 607–631. [CrossRef]
8. Ministerio de Ambiente y Desarrollo Sostenible, y Ministerio de Minas y Energías. *Resolución 40619* **2017** pág 1-3; *Resolución 90963* **2014**, 5.
9. Gutiérrez, O.Y.; Klimova, T. Effect of the support on the high activity of the (Ni)Mo/ZrO_2–SBA-15 catalyst in the simultaneous hydrodesulfurization of DBT and 4,6-DMDBT. *J. Catal.* **2011**, *281*, 50–62. [CrossRef]
10. Rashidi, F.; Sasaki, T.; Rashidi, A.M.; Nemati Kharat, A.; Jozani, K.J. Ultradeep hydrodesulfurization of diesel fuels using highly efficient nanoalumina-supported catalysts: Impact of support, phosphorus, and/or boron on the structure and catalytic activity. *J. Catal.* **2013**, *299*, 321–335. [CrossRef]
11. Chianelli, R.R. Fundamental Studies of Transition Metal Sulfide Hydrodesulfurization Catalysts. *Catal. Rev.* **1984**, *26*, 361–393. [CrossRef]
12. Topsøe, H.; Clausen, B.S.; Massoth, F.E. Hydrotreating Catalysis. In *Catalysis*; Anderson, J.R., Boudart, M., Eds.; Springer: Berlin/Heidelberg, Germany, 1996; pp. 1–269. ISBN 978-3-642-64666-9.
13. Breysse, M.; Portefaix, J.L.; Vrinat, M. Support effects on hydrotreating catalysts. *Catal. Today* **1991**, *10*, 489–505. [CrossRef]
14. Palcheva, R.; Kaluža, L.; Spojakina, A.; Jirátová, K.; Tyuliev, G. NiMo/γ-Al_2O_3 Catalysts from Ni Heteropolyoxomolybdate and Effect of Alumina Modification by B, Co, or Ni. *Chin. J. Catal.* **2012**, *33*, 952–961. [CrossRef]
15. Jirátová, K.; Kraus, M. Effect of support properties on the catalytic activity of HDS catalysts. *Appl. Catal.* **1986**, *27*, 21–29. [CrossRef]
16. Saini, A.R.; Johnson, B.G.; Massoth, F.E. Studies of molybdena—Alumina catalysts XIV. Effect of Cation-Modified Aluminas. *Appl. Catal.* **1988**, *40*, 157–172. [CrossRef]
17. Strohmeier, B. Surface spectroscopic characterization of the interaction between zinc ions and γ-alumina. *J. Catal.* **1984**, *86*, 266–279. [CrossRef]
18. Cabello, C.I.; Botto, I.L.; Thomas, H.J. Anderson type heteropolyoxomolybdates in catalysis:: 1. $(NH_4)_3[CoMo_6O_{24}H_6]\cdot 7H_2O$/γ-Al2O3 as alternative of Co-Mo/γ-Al2O3 hydrotreating catalysts. *Appl. Catal. Gen.* **2000**, *197*, 79–86. [CrossRef]
19. Cabello, C.I.; Cabrerizo, F.M.; Alvarez, A.; Thomas, H.J. Decamolybdodicobaltate(III) heteropolyanion: Structural, spectroscopical, thermal and hydrotreating catalytic properties. *J. Mol. Catal. Chem.* **2002**, *186*, 89–100. [CrossRef]
20. Altamirano, E.; de los Reyes, J.A.; Murrieta, F.; Vrinat, M. Hydrodesulfurization of 4,6-dimethyldibenzothiophene over Co(Ni)MoS2 catalysts supported on alumina: Effect of gallium as an additive. *Catal. Today* **2008**, *133–135*, 292–298. [CrossRef]
21. Díaz de León, J.N.; Picquart, M.; Massin, L.; Vrinat, M.; de los Reyes, J.A. Hydrodesulfurization of sulfur refractory compounds: Effect of gallium as an additive in NiWS/γ-Al_2O_3 catalysts. *J. Mol. Catal. Chem.* **2012**, *363–364*, 311–321. [CrossRef]
22. Cimino, A.; Lo Jacono, M.; Schiavello, M. Effect of zinc, gallium, and germanium ions on the structural and magnetic properties of nickel ions supported on alumina. *J. Phys. Chem.* **1975**, *79*, 243–249. [CrossRef]
23. Zepeda, T.A.; Pawelec, B.; Díaz de León, J.N.; de los Reyes, J.A.; Olivas, A. Effect of gallium loading on the hydrodesulfurization activity of unsupported Ga2S3/WS2 catalysts. *Appl. Catal. B Environ.* **2012**, *111–112*, 10–19. [CrossRef]

24. Petre, A.L.; Auroux, A.; Gervasini, A.; Caldararu, M.; Ionescu, N.I. Calorimetric Characterization of Surface Reactivity of Supported Ga_2O_3 Catalysts. *J. Therm. Anal. Calorim.* **2001**, *64*, 253–260. [CrossRef]
25. Dejonghe, S.; Hubaut, R.; Grimblot, J.; Bonnelle, J.P.; Des Courieres, T.; Faure, D. Hydrometallation of a vanadylporphyrin over sulfided $NiMo\gamma Al_2O_3$, $Mo\gamma Al_2O_3$, and γAl_2O_3 catalysts—Effect of the vanadium deposit on the toluene hydrogenation. *Catal. Today* **1990**, *7*, 569–585. [CrossRef]
26. Rankel, L.; Rollmann, L. Catalytic activity of metals in petroleum and their removal. *Fuel* **1983**, *62*, 44–46. [CrossRef]
27. Lacroix, M.; Boutarfa, N.; Guillard, C.; Vrinat, M.; Breysse, M. Hydrogenating properties of unsupported transition metal sulphides. *J. Catal.* **1989**, *120*, 473–477. [CrossRef]
28. Betancourt, P.; Rives, A.; Scott, C.E.; Hubaut, R. Hydrotreating on mixed vanadium–nickel sulphides. *Catal. Today* **2000**, *57*, 201–207. [CrossRef]
29. Betancourt, P.; Marrero, S.; Pinto-Castilla, S. V–Ni–Mo sulfide supported on Al_2O_3: Preparation, characterization and LCO hydrotreating. *Fuel Process. Technol.* **2013**, *114*, 21–25. [CrossRef]
30. Escalante, Y.; Méndez, F.J.; Díaz, Y.; Inojosa, M.; Morgado, M.; Delgado, M.; Bastardo-González, E.; Brito, J.L. MCM-41-supported vanadium catalysts structurally modified with Al or Zr for thiophene hydrodesulfurization. *Appl. Petrochem. Res.* **2019**, *9*, 47–55. [CrossRef]
31. Ayala-G, M.; Puello, E.; Quintana, P.; González-García, G.; Diaz, C. Comparison between alumina supported catalytic precursors and their application in thiophene hydrodesulfurization: $(NH_4)[NiMo_6O_{24}H_6]\cdot 5H_2O/\gamma\text{-}Al_2O_3$ and $NiMoOx/\gamma\text{-}Al_2O_3$ conventional systems. *RSC Adv.* **2015**, *5*, 102652–102662. [CrossRef]
32. Thommes, M.; Kaneko, K.; Neimark, A.V.; Olivier, J.P.; Rodriguez-Reinoso, F.; Rouquerol, J.; Sing, K.S.W. Physisorption of gases, with special reference to the evaluation of surface area and pore size distribution (IUPAC Technical Report). *Pure Appl. Chem.* **2015**, *87*, 1051–1069. [CrossRef]
33. Sampieri, A.; Pronier, S.; Brunet, S.; Carrier, X.; Louis, C.; Blanchard, J.; Fajerwerg, K.; Breysse, M. Formation of heteropolymolybdates during the preparation of Mo and NiMo HDS catalysts supported on SBA-15: Influence on the dispersion of the active phase and on the HDS activity. *Microporous Mesoporous Mater.* **2010**, *130*, 130–141. [CrossRef]
34. Haneda, M.; Kintaichi, Y.; Shimada, H.; Hamada, H. Selective Reduction of NO with Propene over Ga_2O_3–Al2O3: Effect of Sol–Gel Method on the Catalytic Performance. *J. Catal.* **2000**, *192*, 137–148. [CrossRef]
35. Ueno, A.; Suzuki, H.; Kotera, Y. Particle-size distribution of nickel dispersed on silica and its effects on hydrogenation of propionaldehyde. *J. Chem. Soc. Faraday Trans. 1 Phys. Chem. Condens Phases* **1983**, *79*, 127. [CrossRef]
36. Puello-Polo, E.; Marquez, E.; Brito, J.L. One-pot synthesis of Nb-modified Al_2O_3 support for NiMo hydrodesulfurization catalysts. *J. Sol-Gel Sci. Technol.* **2018**, *88*, 90–99. [CrossRef]
37. International Centre for Diffraction Data®(ICDD®). *Power Diffraction File*; ICDD: Newtown Square, PA, USA, 1995.
38. Galtayries, A.; Wisniewski, S.; Grimblot, J. Formation of thin oxide and sulphide films on polycrystalline molybdenum foils: Characterization by XPS and surface potential variations. *J. Electron Spectrosc. Relat. Phenom.* **1997**, *87*, 31–44. [CrossRef]
39. Weber, T.; Muijsers, J.C.; van Wolput, J.H.M.C.; Verhagen, C.P.J.; Niemantsverdriet, J.W. Basic Reaction Steps in the Sulfidation of Crystalline MoO_3 to MoS_2, As Studied by X-ray Photoelectron and Infrared Emission Spectroscopy. *J. Phys. Chem.* **1996**, *100*, 14144–14150. [CrossRef]
40. Aigler, J.M.; Brito, J.L.; Leach, P.A.; Houalla, M.; Proctor, A.; Cooper, N.J.; Hall, W.K.; Hercules, D.M. ESCA study of "model" allyl-based molybdenum/silica catalysts. *J. Phys. Chem.* **1993**, *97*, 5699–5702. [CrossRef]
41. Le, Z.; Afanasiev, P.; Li, D.; Long, X.; Vrinat, M. Solution synthesis of the unsupported Ni–W sulfide hydrotreating catalysts. *Catal. Today* **2008**, *130*, 24–31. [CrossRef]
42. Wang, X.; Ozkan, U.S. Characterization of Active Sites over Reduced $Ni–Mo/Al_2O_3$ Catalysts for Hydrogenation of Linear Aldehydes. *J. Phys. Chem. B* **2005**, *109*, 1882–1890. [CrossRef]
43. Schön, G. Auger and direct electron spectra in X-ray photoelectron studies of zinc, zinc oxide, gallium and gallium oxide. *J. Electron Spectrosc. Relat. Phenom.* **1973**, *2*, 75–86. [CrossRef]
44. Escaño, M.C.S.; Asubar, J.T.; Yatabe, Z.; David, M.Y.; Uenuma, M.; Tokuda, H.; Uraoka, Y.; Kuzuhara, M.; Tani, M. On the presence of Ga2O sub-oxide in high-pressure water vapor annealed AlGaN surface by combined XPS and first-principles methods. *Appl. Surf. Sci.* **2019**, *481*, 1120–1126. [CrossRef]

45. Rakmae, S.; Osakoo, N.; Pimsuta, M.; Deekamwong, K.; Keawkumay, C.; Butburee, T.; Faungnawakij, K.; Geantet, C.; Prayoonpokarach, S.; Wittayakun, J.; et al. Defining nickel phosphides supported on sodium mordenite for hydrodeoxygenation of palm oil. *Fuel Process. Technol.* **2020**, *198*, 106236. [CrossRef]
46. Li, M.; Li, H.; Jiang, F.; Chu, Y.; Nie, H. The relation between morphology of (Co)MoS$_2$ phases and selective hydrodesulfurization for CoMo catalysts. *Catal. Today* **2010**, *149*, 35–39. [CrossRef]
47. Liu, H.; Liu, C.; Yin, C.; Liu, B.; Li, X.; Li, Y.; Chai, Y.; Liu, Y. Low temperature catalytic hydrogenation naphthalene to decalin over highly-loaded NiMo, NiW and NiMoW catalysts. *Catal. Today* **2016**, *276*, 46–54. [CrossRef]
48. Barrett, E.P.; Joyner, L.G.; Halenda, P.P. The Determination of Pore Volume and Area Distributions in Porous Substances. I. Computations from Nitrogen Isotherms. *J. Am. Chem. Soc.* **1951**, *73*, 373–380. [CrossRef]
49. Farojr, A.; Dossantos, A. Cumene hydrocracking and thiophene HDS on niobia-supported Ni, Mo and Ni–Mo catalysts. *Catal. Today* **2006**, *118*, 402–409. [CrossRef]
50. Froment, G.F.; De Wilde, J.; Bischoff, K.B. *Chemical Reactor Analysis and Design*, 3rd ed.; Wiley: Hoboken, NJ, USA, 2011; ISBN 978-0-470-56541-4.
51. Moulijn, J.A.; Tarfaoui, A.; Kapteijn, F. General aspects of catalyst testing. *Catal. Today* **1991**, *11*, 1–12. [CrossRef]
52. Farag, H. Kinetic Analysis of the Hydrodesulfurization of Dibenzothiophene: Approach Solution to the Reaction Network. *Energy Fuels* **2006**, *20*, 1815–1821. [CrossRef]
53. Vargas-Villagrán, H.; Ramírez-Suárez, D.; Ramírez-Muñoz, G.; Calzada, L.A.; González-García, G.; Klimova, y.T.E. Tuning of activity and selectivity of Ni/(Al)SBA-15 catalysts in naphthalene hydrogenation. *Catal. Today* **2019**, S0920586119305103. [CrossRef]

© 2020 by the authors. Licensee MDPI, Basel, Switzerland. This article is an open access article distributed under the terms and conditions of the Creative Commons Attribution (CC BY) license (http://creativecommons.org/licenses/by/4.0/).

Article

Montmorillonite K10: An Efficient Organo-Heterogeneous Catalyst for Synthesis of Benzimidazole Derivatives

Sonia Bonacci, Giuseppe Iriti, Stefano Mancuso, Paolo Novelli, Rosina Paonessa, Sofia Tallarico and Monica Nardi *

Dipartimento di Scienze della Salute, Università Magna Græcia, Viale Europa, Germaneto, 88100 Catanzaro, Italy; s.bonacci@unicz.it (S.B.); giuseppeiriti94@gmail.com (G.I.); stefanoman27@gmail.com (S.M.); paolo.novelli92@gmail.com (P.N.); r.paonessa@unicz.it (R.P.); sofia.tallarico@outlook.it (S.T.)
* Correspondence: monica.nardi@unicz.it; Tel.: +39-0961-3694116

Received: 6 July 2020; Accepted: 25 July 2020; Published: 28 July 2020

Abstract: The use of toxic solvents, high energy consumption, the production of waste and the application of traditional processes that do not follow the principles of green chemistry are problems for the pharmaceutical industry. The organic synthesis of chemical structures that represent the starting point for obtaining active pharmacological compounds, such as benzimidazole derivatives, has become a focal point in chemistry. Benzimidazole derivatives have found very strong applications in medicine. Their synthesis is often based on methods that are not convenient and not very respectful of the environment. A simple montmorillonite K10 (MK10) catalyzed method for the synthesis of benzimidazole derivatives has been developed. The use of MK10 for heterogeneous catalysis provides various advantages: the reaction yields are decidedly high, the work-up procedures of the reaction are easy and suitable, there is an increase in selectivity and the possibility of recycling the catalyst without waste formation is demonstrated. The reactions were carried out in solvent-free conditions and in a short reaction time using inexpensive and environmentally friendly heterogeneous catalysis. It has been shown that the reaction process is applicable in the industrial field.

Keywords: heterogeneous catalysis; montmorillonite; benzimidazoles

1. Introduction

Benzimidazole is a hetero bicyclic aromatic organic compound consisting in the fusion of benzene and imidazole. The benzimidazole ring is very well known in nature thanks to its various therapeutic applications. Its "nucleus" is present in many important molecules such as, for example, vitamin B_{12} [1].

In the early nineties, various benzimidazole derivatives were synthesized, obtaining fluorine, propylene and tetrahydroquinoline derivatives with greater stability and biological activity [2,3], while derivatives with an electron-donating group have proven to have good antiulcer activity [4,5], such as omeoprazole.

Recently, the therapeutic effects of benzimidazole derivatives in diseases such as ischemia-reperfusion injury or hypertension have been demonstrated [6].

Thanks to their various pharmacological properties, various synthetic methodologies have been developed in the field of organic synthesis.

The first synthetic methodologies reported in the literature are based on the reaction between o-phenylenediamine and carboxylic acids or their derivatives [7,8].

Subsequently, the reaction process was made easier by replacing the carboxylic acids with aldehydes, obtaining 2-substituted and 1,2-substituted benzimidazole derivatives. Numerous methods are reported for the condensation of substituted *o*-phenylenediamine with aldehydes catalyzed by metal triflate such as Sc(OTf)$_3$ or Yb (OTf)$_3$ [9], TiCl$_3$OTf [10], different oxidizing agents [11–14] and lanthanides such as Lewis acid catalysts [15,16]. However, these protocols present several problems that make the methods less convenient due to long reaction times and the use of expensive reagents and toxic organic solvents. Furthermore, non-recoverable, difficult to prepare and poorly selective catalysts are often used [17–22].

Since the development of new synthetic methods to produce potential drug compounds has always played a relevant role in scientific research, in recent years, the use of recyclable heterogeneous catalysts has become very important. Their use is favored because of their particularly versatile properties, low cost and thermal stability. In addition, reactions catalyzed by solid supports or in a solid state provide better selectivity in the products, compared to solution phase reactions.

These heterogeneous catalysts have found widespread application in eco-sustainable organic synthesis, showing higher activity than homogeneous catalysts [23,24]. Their use in the pharmaceutical industry is favored because of their easy recovery and stability and their ability to minimize waste. The synthesis of Lewis acid heterogeneous catalysts from waste materials has become increasingly popular over recent years [25], such as in the case of sulfonic-acid-functionalized activated carbon prepared from matured tea leaf, tested for synthesis of 2-substituted benzimidazole and benzothiazole [26].

The use of toxic solvents in the pharmaceutical industry is a serious problem for the environment and human health, but in recent years, green chemistry principles have influenced the activities of the drug industry, introducing less use of classic organic solvents [27–30], cuts in waste production with the use of recyclable reagents [31–35] and the use of environmental organic synthetic methods.

Various research studies have been conducted on the use of "green" solvents [36], principally bio-solvents [37–42], ionic liquids [43–45], deep eutectic solvents [46–51], supercritical fluids [52,53] or water [54–62]. Certainly, the use of experimental methods based on solvent-free or solid state reaction conditions may reduce pollution. Green reactions may be also carried out using the reactants alone. Often the same reactions involve the use of solid supports (clays, zeolites, silica, alumina or other matrices), easing the experimental and work-up procedures, improving yields, increasing the reaction rate and considerably lowering the environmental impact [63–65]. In this context, therefore, solid Lewis acid catalysts are widely used and thermal process [66,67] can be employed to lead the reactions.

The use of microwaves (MW) in solvent-free reactions [68–71] has been particularly important for industrial production. MW irradiation increases the rate of chemical reactions, thus showing great potential in innovative chemical reaction processes [72]. This improvement is particularly demonstrated in heterogeneous catalytic systems, compared with conventional heating under identical temperature conditions, presumably due to interaction(s) between the MW radiation fields and the catalyst itself. For the above reason, it has given rise, over the years, to a strong interest in the field of the synthesis of pharmaceutical compounds [73–83].

In this regard, montmorillonite represents an ideal heterogeneous eco-sustainable catalyst thanks to its low cost, ease of handling, easy recovery by filtration method and possibility of use in chemical reactions in solvent-free conditions under microwaves or ultrasound irradiation [84]. Like other clay catalysts, it is widely available and has a high surface area containing both Brønsted and Lewis acid sites catalyzing organic reactions [85–88].

Recently, a simple and eco-friendly protocol for the synthesis of some novel substituted 2-arylbenzimidazoles was developed using ZrOCl$_2$·nH$_2$O supported on montmorillonite K10 [89,90]. The synthetic process involves only the formation of the 2-benzimidazole derivative, and requires the preparation of a catalyst and the use of water as a solvent. Moreover, acid treated modified montmorillonite clay was used as a catalyst precursor for the synthesis of benzimidazoles, but the pretreatment of the catalyst and the use of toluene as a solvent makes the synthetic process unsustainable [91]. Other zeolites have been tested for the synthesis of benzimidazoles, but the experimental procedures do not show selectivity [92–94].

Considering the stability, catalytic activity and selectivity of MK10 tested in the synthesis reactions of bifunctionalized cyclopentenones [95] and our experience in developing environmental reactions for the synthesis of pharmaceutical azo-compounds [96–100], we present a new and selective synthetic method to obtain benzimidazole derivatives in a solvent-free reaction, testing MK10 as a heterogeneous catalyst.

2. Results

In our preliminary experiment, we choose o-phenylenediamine, o-PDA, (1 mmol) and benzaldehyde as starting materials to selectively obtain 1,2-disubstituted benzimidazole derivative **1a** (Table 1).

Table 1. Optimization of the reaction conditions. [a]

Entry	MK10 wt (%) [b]	Molar Ratio o-PDA: Benzaldehyde	Temp (°C)	Time (min)	Conversion (%) [c]	Selectivity (%) [d]
1	10	1:1	rt	120	19.3	12.0
2	10	1:2	rt	120	20.9	53.0
3	10	1:2	60	120	79.6	65.1
4	10	1:1	80	120	80.9	33.3
5	10	1:1	100	60	99.9	38.3
6	10	1:2	100	60	99.9	75.0
7	-	1:2	100	90	45.0	49.0
8 [e]	20	1:1	60	5	99.9	18.2
9 [e]	20	1:2	60	5	99.9	98.5

[a] General reaction conditions: o-PDA (1 mmol) and benzaldehyde (1 or 2 mmol) were stirred for 5–120 min at different temperatures and different wt (%) of MK10. [b] wt % with respect to amine. [c] Percent conversion of the o-PDA calculated from GC/MS data. [d] Percent yield calculated from GC/MS data of the corresponding disubstituted benzimidazole derivative. By-product obtained is constituted by 2-phenyl-benzimidazole (**1b**). [e] Reaction mixture under MW irradiation; the temperature was controlled in the microwave reactor.

Initially, we tested the effect of MK10 on the model reaction by performing the reaction (Table 1, entry 1) using 10 wt% of MK10 with respect to o-phenylendiamine. The reaction mixture, stirred at room temperature for 2 h, consists of diamine and benzaldehyde in a 1:1 and 1:2 molar ratio, respectively (Table 1, entries 1 and 2). The reaction is monitored by thin layer chromatography (TLC) and gas chromatography/mass spectrometry (GC/MS) analysis.

The GC/MS analysis showed the low conversion of the reagents within 120 min and low selectivity even when using 2 mmol benzaldehyde (Table 1, entry 2). At the higher temperature, 60 °C, the 1,2-disubstituted benzimidazole derivative **1a** was favored (65.1% yields), but the 2-substituted benzimidazole derivative **1b** in 34.9% yields was also obtained (Table 1, entry 3), thus not improving selectivity. The selectivity was worsened using 1 mmol benzaldehyde at 80 °C (Table 1, entry 4). The GC-MS analysis showed the presence of the corresponding 2-phenyl-benzimidazole by-product (66.7% yield) in 2 h. The model reaction showed the complete conversion of o-phenilendiammine when the same reaction was performed at higher temperatures (100 °C) (Table 1, entries 5 and 6) in 1 h. By increasing the molar ratio of benzaldeide (2 mmol) at the same temperature in the same reaction time (60 min), a better selectivity was observed (Table 1, entry 6). When the same reaction was carried out in the absence of a catalyst and in a longer reaction time (90 min), no complete conversion of o-PDA and more by-product formation were observed (Table 1, entry 7).

We obtained the complete conversion when the amount of catalyst was increased to 20 wt% of MK10 at 60 °C under MW irradiation (Table 1, entry 8), achieving 2-phenyl-benzimidazole as the principal product (81.2% yield) and using 1 mmol of benzaldehyde. Surprisingly, we gained the desired product, 1-benzyl-2-phenyl-benzimidazole **1a**, in 98.5% yield and in only 5 min at 60 °C (Table 1, entry 9) using 2 mmol of benzaldehyde.

The use of the heterogeneous catalyst has made the reaction process even more eco-sustainable than the previously developed methodologies, in terms of both faster reaction times and greater selectivity of product formation.

The fundamental contribution that a heterogeneous catalyst makes to the sustainability of a reaction process is its being recyclable.

To demonstrate this, after testing MK10 in the reaction model system using the best reaction conditions (Table 1, entry 9), the final reaction mixture was treated with ethyl acetate. The MK10 was recovered from the organic solution by filtration, washed with ethyl acetate (3 mL) four times and dried in an oven (40 °C). The combined organic phases were concentrated by vacuum rotary evaporation.

The percent conversion and selectivity were analyzed by GC/MS. The recovered catalyst was used directly for the next run, adding new, fresh reagents following the procedures reported in the literature [91] (Figure 1).

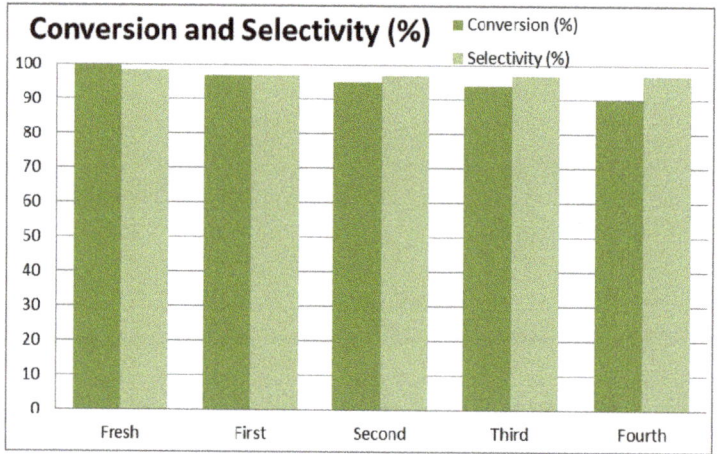

Figure 1. Cycling performance of MK10 in synthesis of 1-benzyl-2-phenyl-benzimidazole **1a** under MW irradiation.

In order to demonstrate the potential industrial applicability as a green procedure, the model reaction was tested on a large scale using 10 mmol of o-phenylendiammine, 20 mmol of benzaldehyde and the respective amount of MK10. The reaction was completed in 25 min with excellent yield (95%) after simple extraction with ethyl acetate.

The experimental method was applied using o-PDA and different aldehydes to obtain 1,2-disubstituted benzimidazole derivatives. Quantitative yields superior to 90% were obtained in cases of aldehydes containing electron-donor groups (Table 2, entries 1–3 and entries 6 and 7).

The reactions performed with aldehydes containing electron-withdrawing groups such as p-chloro or p-nitro benzaldehyde (Table 2, entries 4 and 5) did not afford the disubstituted derivative, but did afford the corresponding 2-monosubstituted benzimidazoles (**4b** and **5b**) in good yields (detected by GC/MS). In this case, the monosubstituted product can be separated from the excess of benzaldehyde through chromatographic separation.

The same reactions performed using 1 molar amount of aldehydes afforded the corresponding 2-monosubstituted benzimidazoles (**1b–8b**) in good yields demonstrating, once again, the selectivity of the adopted reaction process (Table 3). This result was in accordance with the data reported in the literature [50,99].

Table 2. Synthesis of 1,2-disubstituted benzimidazoles. [a]

Entry	Aldehyde	Product	Conversion (%)	Yield (%) [b]
1	benzaldehyde	1a	99.9	95.0
2	4-methylbenzaldehyde	2a	98.7	96.6
3	4-methoxybenzaldehyde	3a	99.9	99.6
4 [c]	4-chlorobenzaldehyde	4a	982	0
5 [c]	4-nitrobenzaldehyde	5a	97.3	0
6	propanal	6a	91.0	90.8
7	acetaldehyde	7a	97.8	95.1
8	phenylacetone	8a	96.8	93.8

[a] General reaction conditions: 1 mmol of o-OPD and 2 mmol of aldehyde are added to 20% mw to amine of MK10. The reaction was conducted in a Synthos 3000 microwave oven (Anton-Paar) at 60 °C for 5 min. The reaction mixture was then washed with AcOEt (3 × 3 mL) and filtered to obtain MK10. The combined organic phases were dried over Na_2SO_4, filtered and evaporated under reduced pressure to give the corresponding products 1a–8a. [b] Percent yield calculated from GC/MS data. The corresponding 1,2-disubstituted benzimidazole derivative was recovered as the only product. [c] Product a was not detected. Only the corresponding 2-substituted derivatives, 4b and 5b (Conversion 98% and 97%, respectively, of o-OPD calculated from GC/MS) were detected by GC/MS.

Table 3. Synthesis of 2-monosubstituted benzimidazoles. [a]

Entry	Aldehyde	Product	Conversion (%)	Yield (%) [b]
1	benzaldehyde	1b	99.9	95.0
2	4-methylbenzaldehyde	2b	95.9	97.8
3	4-methoxybenzaldehyde	3b	99.9	99.0
4	4-chlorobenzaldehyde	4b	90.6	98.3
5	4-nitrobenzaldehyde	5b	89.6	97.3
6	propanal	6b	91.0	90.8
7	acetaldehyde	7b	97.8	94.8
8	phenylacetone	8b	96.8	94.1

[a] General reaction conditions: 1 mmol of o-PDA and 1 mmol of aldehyde are added to 20% mw to amine of MK10. The reaction was conducted in a Synthos 3000 microwave oven (Anton-Paar) at 60 °C for 5 min. The corresponding products, monosubstituted benzimidazoles **1b–8b**, were isolated as previously described to obtain disubstituted benzimidazoles (Table 2, footnote a). [b] Percent yield calculated from GC/MS data.

In conclusion, in the development of a green procedure, the recyclability of the heterogeneous catalyst MK10 is an essential feature. All reactions were performed in short reaction times (5 min) and with reaction yields of 90% to 99% (Tables 2 and 3).

Unlike the reaction procedures reported in the literature, the described method does not require the use of solvents [99] or the synthesis of deep eutectic solvents [50] essential to perform the complete reaction process. The proposed method reduces energy consumption and reaction time, making the process industrially acceptable.

3. Materials and Methods

3.1. General Methods

Montmorillonite K10 clay and all chemical reagents were obtained from Sigma-Aldrich. The chemical composition (wt%) of the clay (main elements) was SiO_2: 67.6; Al_2O_3: 14.6; Fe_2O_3: 2.9; MgO: 1.8.

All reactions were monitored by a GC-MS Shimadzu workstation. It is constituted by a GC 2010 (equipped with a 30 m QUADREX 007-5MS capillary column, operating in the "split" mode, 1 mL min^{-1} flow of He as carrier gas, (Shimadzu Corporation, Kyoto, Japan).

^{1}H-NMR and ^{13}C-NMR spectra were recorded at 300 MHz and at 75 MHz, respectively, using a Bruker WM 300 system, (Bruker Corporation, Massachusetts, USA). The samples were solubilized in CDCl$_3$ using tetramethylsilane (TMS) as a reference (δ 0.00). Chemical shifts are given in parts per million (ppm), and coupling constants (J) are given in hertz. For ^{13}C-NMR, the chemical shifts are relative to CDCl$_3$ (δ 77.0).

A Synthos 3000 instrument from Anton Paar, (Minoh City, Osaka, Japan), equipped with a 4 × 24MG5 rotor, was used for the MW-assisted reactions. An external IR sensor monitored the temperature at the base of each reaction vessel.

3.2. General Procedure for the Synthesis of 1,2-Substituted Benzimidazoles **1a–8a**

The aldehyde (2 mmol) was added to the o-PDA (1 mmol) and MK10 (20 mg). The obtained mixture was reacted for 5 min under microwave heating, at a temperature of 60 °C (IR limit). After complete conversion of o-phenilendiammine, the MK10 was separated from the reaction mixture by filtration and washed with ethyl acetate (4 × 3 mL). The products were isolated after evaporation of the solvent to afford compounds in 90–99% yields. The NMR spectral data were in accordance with those reported in the literature [50] (See Supplementary Materials).

3.3. General Procedure for the Synthesis of 2-Substituted Benzimidazoles **1b–8b**

The synthesis procedure of the mono-substituted imidazoles derived was carried out under the same conditions used for the synthesis of the 1,2-substituted benzimidazoles. In this case, however, the aldehydes were used in an amount equal to 1mmol. After complete conversion of o-PDA in the 2-monosubstituted benzimidazoles (5 min), the products were isolated as previously described. The NMR spectral data were in accordance with those reported in the literature [50] (See Supplementary Materials).

3.4. Catalyst Recycling

The MK10 was separated from the reaction mixture by rapid filtration, then washed with ethyl acetate (3 mL) four times and dried in an oven (50 °C).

4. Conclusions

A fast, cheap, simple and environmentally sustainable method has been developed for the synthesis of 1,2-bisubstituted benzimidazoles and 2-substituted benzimidazoles. Microwave assistance was crucial to obtain the products in only five minutes.

Moreover, this proposed method produces very low quantities of reaction waste. MK10 was recycled and reused for four consecutive cycles without any significant loss in catalytic activity, as previously demonstrated [92].

Furthermore, compared to recently reported procedures, the proposed method does not require a previous treatment for the preparation of deep eutectic solvents (DESs) as eco-friendly and sustainable solvent and catalytic systems (the procedure of preparation of DESs requires 2 h at 80 °C), necessary to perform the subsequent synthesis reaction of benzimidazoles [50].

All this means that the use of the heterogeneous catalyst MK10 provides a synthetic procedure that considerably reduces reaction times and energy costs, further promoting industrial application.

Supplementary Materials: The following are available online at http://www.mdpi.com/2073-4344/10/8/845/s1. Experimental Section, General Procedure for the Synthesis of 1,2-Substituted Benzimidazoles 1a–8a, General Procedure for the Synthesis of 2-Substituted Benzimidazoles 1b–8b, Catalyst recycling, ^{1}H NMR and ^{13}C NMR of compounds 1a–3a, 6a–8a, ^{1}H NMR and ^{13}C NMR of compounds 1b–8b.

Author Contributions: M.N. conceived and designed the experiments; S.B. performed the experiments; G.I., S.M., P.N. and S.T. analyzed the data; S.B. and R.P. wrote the paper. All authors have read and agreed to the published version of the manuscript.

Funding: This research received funding from Dipartimento di Scienze della Salute, Università Magna Græcia, Italy.

Conflicts of Interest: The authors declare no conflict of interest.

References

1. Emerson, G.; Brink, N.G.; Holly, F.W.; Koniuszy, F.; Heyl, D.; Folker, K. Vitamin B_{12}. VIII. Vitamin B_{12}-Like Activity of 5,6-Dimethylbenzimidazole and Tests on related compounds. *J. Am. Chem. Soc.* **1950**, *72*, 3084–3085. [CrossRef]
2. Kubo, K.; Oda, K.; Kaneko, T.; Satoh, H.; Nohara, A. Synthesis of 2-(4- Fluoroalkoxy-2-pyridyl) methyl] sulfinyl]-1H-benzimidazoles as Antiulcer Agents. *Chem. Pharm. Bull.* **1990**, *38*, 2853–2858. [CrossRef] [PubMed]
3. Uchida, M.; Chihiro, M.; Morita, S.; Yamashita, H.; Yamasaki, K.; Kanbe, T.; Yabuuchi, Y.; Nakagawz, K. Synthesis and Antiulcer Activity of 4- Substituted 8-[(2-Benzimidazolyl) sulfinylmethyl]-1, 2, 3, 4-tetrahydroquinolines and Related Compounds. *Chem. Pharm. Bull.* **1990**, *38*, 1575–1586. [CrossRef] [PubMed]
4. Grassi, A.; Ippen, J.; Bruno, M.; Thomas, G.; Bay, P. A thiazolylamino benzimidazole derivative with gastroprotective properties in the rat. *Eur. J. Pharmacol.* **1991**, *195*, 251–259. [CrossRef]
5. Ozkay, Y.; Tunali, Y.; Karaca, H.; Isikdag, I. Antimicrobial activity and a SAR study of some novel benzimidazole derivatives bearing hydrazones moiety. *Eur. J. Med. Chem.* **2010**, *45*, 3293–3298. [CrossRef]
6. Algul, O.; Karabulut, A.; Canacankatan, N.; Gorur, A.; Sucu, N.; Vezir, O. Apoptotic and anti-angiogenic effects of benzimidazole compounds: Relationship with oxidative stress mediated ischemia/reperfusion injury in rat hind limb. *Antiinflamm Antiallergy Agents Med. Chem.* **2012**, *11*, 267–275. [CrossRef]
7. Thakuria, H.; Das, G. An expeditious one-pot solvent-free synthesis of benzimidazole derivatives. *ARKIVOC* **2008**, *15*, 321–328.
8. Rithe, S.R.; Jagtap, R.S.; Ubarhande, S.S. One Pot Synthesis of Substituted Benzimidazole Derivatives And Their Characterization. *RASAYAN J. Chem.* **2015**, *8*, 213–217.
9. Liyan, F.; Wen, C.; Lulu, K. Highly chemoselective synthesis of benzimidazoles in Sc(OTf)3-catalyzed system. *Heterocycles* **2015**, *91*, 2306–2314.
10. Bahrami, K.; Khodaei, M.M.; Kavianinia, I. H_2O_2/HCl as a new and efficient system for synthesis of 2-substituted benzimidazoles. *J. Chem. Res.* **2006**, *12*, 783–784. [CrossRef]
11. Ma, H.; Han, X.; Wang, Y.; Wang, J. A simple and efficient method for synthesis of benzimidazoles using FeBr3 or Fe(NO3)3·9H2O as catalyst. *ChemInform* **2007**, *38*, 1821–1825. [CrossRef]
12. Du, L.-H.; Wang, Y.-G. A rapid and efficient synthesis of benzimidazoles using hypervalent iodine as oxidant. *Synthesis* **2007**, *5*, 675–678.
13. Sontakke, V.A.; Ghosh, S.; Lawande, P.P.; Chopade, B.A.; Shinde, V.S. A simple, efficient synthesis of 2-aryl benzimidazoles using silica supported periodic acid catalyst and evaluation of anticancer activity. *ISRN Org. Chem.* **2013**, *2013*, 1–7. [CrossRef] [PubMed]
14. Kumar, K.R.; Satyanarayana, P.V.V.; Reddy, B.S. $NaHSO_4$-SiO_2 promoted synthesis of benzimidazole derivatives. *Arch. Appl. Sci. Res.* **2012**, *4*, 1517–1521.
15. Venkateswarlu, Y.; Kumar, S.R.; Leelavathi, P. Facile and efficient one-pot synthesis of benzimidazoles using lanthanum chloride. *Org. Med. Chem. Lett.* **2013**, *3*, 2–8.
16. Martins, G.M.; Puccinelli, T.; Gariani, R.A.; Xavier, F.R.; Silveira, C.C.; Mendes, S.R. Facile and efficient aerobic one-pot synthesis of benzimidazoles using $Ce(NO_3)_3 \cdot 6H_2O$ as promoter. *Tetrahedron Lett.* **2017**, *58*, 1969–1972. [CrossRef]
17. Mobinikhaledi, A.; Hamta, A.; Kalhor, M.; Shariatzadeh, M. Simple Synthesis and Biological Evaluation of Some Benzimidazoles Using Sodium Hexafluroaluminate, Na3 AlF6, as an Efficient Catalyst. *Iran. J. Pharm. Res.* **2014**, *13*, 95–101.
18. Birajdar, S.S.; Hatnapure, G.D.; Keche, A.P.; Kamble, V.M. Synthesis of 2-substituted-1 H-benzo[d]imidazoles through oxidative cyclization of O-phenylenediamine and substituted aldehydes using dioxanedibromide. *Res. J. Pharm. Biol. Chem. Sci.* **2014**, *5*, 487–493.
19. Srinivasulu, R.; Kumar, K.R.; Satyanarayana, P.V.V. Facile and Efficient Method for Synthesis of Benzimidazole Derivatives Catalyzed by Zinc Triflate. *Green Sustain. Chem.* **2014**, *4*, 33–37. [CrossRef]
20. Sehyun, P.; Jaehun, J.; Eun, J.C. Visible-Light-Promoted Synthesis of Benzimidazoles. *J. Org. Chem.* **2014**, *352*, 4148–4154.
21. Vishvanath, D.P.; Ketan, P.P. Synthesis of Benzimidazole and Benzoxazole Derivatives Catalyzed by Nickel Acetate as Organometallic Catalyst. *Int. J. ChemTech Res.* **2014**, *8*, 457–465.

22. Procopio, A.; De Nino, A.; Nardi, M.; Oliverio, M.; Paonessa, R.; Pasceri, R. A New Microwave-Assisted Organocatalytic Solvent-Free Synthesis of Optically Enriched Michael Adducts. *Synlett* **2010**, *12*, 1849–1853. [CrossRef]
23. Deng, Q.; Wang, R. Heterogeneous MOF catalysts for the synthesis of trans-4,5-diaminocyclopent-2-enones from furfural and secondary amines. *Catal. Commun.* **2019**, *120*, 11–16.
24. Thomas, J.M.; Raja, R.; Lewis, D.W. Single-site heterogeneous catalysts. *Angew. Chem. Int. Ed.* **2005**, *44*, 6456. [CrossRef] [PubMed]
25. Osman, A.I.; Abu-Dahrieh, J.K.; McLaren, M.; Laffir, F.; Rooney, D.W. Characterisation of Robust Combustion Catalyst from Aluminium Foil Waste. *ChemistrySelect* **2018**, *3*, 1545–1550. [CrossRef]
26. Goswami, M.; Dutta, M.M.; Phukan, P. Sulfonic-acid-functionalized activated carbon made from tea leaves as green catalyst for synthesis of 2-substituted benzimidazole and benzothiazole. *Res. Chem. Intermed.* **2018**, *44*, 1597–1615. [CrossRef]
27. Nelso, W.M. *Green Solvents for Chemistry Perspectives and Practice*; Oxford University Press: Oxford, UK, 2004.
28. Mikami, K. *Green Reaction Media in Organic Synthesis*; Blackwell: Tokyo, Japan, 2005.
29. Clark, J.H.; Tavener, S.J. Alternative Solvents: Shades of Green. *Org. Process Res. Dev.* **2007**, *11*, 149–155. [CrossRef]
30. Ballini, R.; Bosica, G.; Carloni, L.; Maggi, R.; Sartori, G. Zeolite HSZ-360 as a new reusable catalyst for the direct acetylation of alcohols and phenols under solventless conditions. *Tetrahedron Lett.* **1998**, *39*, 6049–6052. [CrossRef]
31. Bartoli, G.; Dalpozzo, R.; De Nino, A.; Maiuolo, L.; Nardi, M.; Procopio, A.; Tagarelli, A. Cerium(III) Triflate versus Cerium(III) Chloride: Anion Dependence of Lewis Acid Behavior in the Deprotection of PMB Ethers. *Eur. J. Org. Chem.* **2004**, *10*, 2176–2180. [CrossRef]
32. Procopio, A.; Cravotto, G.; Oliverio, M.; Costanzo, P.; Nardi, M.; Paonessa, R. An Eco-Sustainable Erbium(III)-Catalysed Method for Formation/Cleavage of O-tert-butoxy carbonates. *Green Chem.* **2011**, *13*, 436–443. [CrossRef]
33. Oliverio, M.; Costanzo, P.; Macario, A.; De Luca, G.; Nardi, M.; Procopio, A. A Bifuctional Heterogeneous Catalyst Erbium-Based: A Cooperative Route Towards C-C Bond Formation. *Molecules* **2014**, *19*, 10218–10229. [CrossRef] [PubMed]
34. Procopio, A.; Das, G.; Nardi, M.; Oliverio, M.; Pasqua, L. A Mesoporous Er(III)-MCM-41 Catalyst for the Cyanosilylation of Aldehydes and Ketones under Solvent-free Conditions. *ChemSusChem* **2008**, *1*, 916–919. [CrossRef] [PubMed]
35. Procopio, A.; Costanzo, P.; Curini, M.; Nardi, M.; Oliverio, M.; Sindona, G. Erbium(III) Chloride in Ethyl Lactate as a Smart Ecofriendly System for Efficient and Rapid Stereoselective Synthesis of trans-4,5-Diaminocyclopent-2-enones. *ACS Sustain. Chem. Eng.* **2013**, *1*, 541–544. [CrossRef]
36. Virot, M.; Tomao, V.; Ginies, C.; Chemat, F. Total lipid extraction of food using d-limonene as an alternative to n-hexane. *Chromatographia* **2008**, *68*, 311–313. [CrossRef]
37. Lapkin, A.; Plucinski, P.K.; Cutler, M. Comparative assessment of technologies for extraction of artemisinin. *J. Nat. Prod.* **2006**, *69*, 1653–1664. [CrossRef]
38. Pereira, C.S.M.; Silva, V.M.T.M.; Rodrigues, A.E. Ethyl lactate as a solvent: Properties, applications and production processes. *Green Chem.* **2011**, *13*, 2658–2671. [CrossRef]
39. García, J.I.; García-Marín, H.; Pires, E. Glycerol based solvents: Synthesis, properties and applications. *Green Chem.* **2014**, *16*, 1007–1033. [CrossRef]
40. Nardi, M.; Oliverio, M.; Costanzo, P.; Sindona, G.; Procopio, A. Eco-friendly stereoselective reduction of α,β-unsaturated carbonyl compounds by Er(OTf)$_3$/NaBH$_4$ in 2-MeTHF. *Tetrahedron* **2015**, *71*, 1132–1135. [CrossRef]
41. Nardi, M.; Herrera Cano, N.; De Nino, A.; Di Gioia, M.L.; Maiuolo, L.; Oliverio, M.; Santiago, A.; Sorrentino, D.; Procopio, A. An eco-friendly tandem tosylation/Ferrier N-glycosylation of amines catalyzed by Er(OTf)$_3$ in 2-MeTHF. *Tetrahedron Lett.* **2017**, *58*, 1721–1726. [CrossRef]
42. Weishi Miao, W.; Chan, T.H. Ionic-Liquid-Supported Synthesis: A Novel Liquid-Phase Strategy for Organic Synthesis. *Acc. Chem. Res.* **2006**, *39*, 897–908.
43. Abbott, A.P.; Davies, D.L.; Capper, G.; Rasheed, R.K.; Tambyrajah, V. Ionic Liquids and Their Use As solvents. U.S. Patent 7,183,433, 27 February 2007.
44. Di Gioia, M.L.; Costanzo, P.; De Nino, A.; Maiuolo, L.; Nardi, M.; Olivito, F.; Procopio, A. Simple and efficient Fmoc removal in ionic liquid. *RSC Adv.* **2017**, *7*, 36482–36491. [CrossRef]

45. De Nino, A.; Maiuolo, L.; Merino, P.; Nardi, M.; Procopio, A.; Roca-López, D.; Russo, B.; Algieri, V. Efficient organocatalyst supported on a simple ionic liquid as a recoverable system for the asymmetric diels-alder reaction in the presence of water. *ChemCatChem* **2015**, *7*, 830–835. [CrossRef]
46. Abbott, A.P.; Capper, G.; Davies, D.L.; Rasheed, R.K.; Tambyrajah, V. Novel solvent properties of choline chloride/urea mixtures. *Chem. Commun.* **2003**, *1*, 70–71. [CrossRef] [PubMed]
47. Gorke, J.T.; Srienc, F.; Kazlauskas, R.J. Hydrolase-catalyzed biotransformations in deep eutectic solvents. *Chem. Commun.* **2008**, *10*, 1235–1237. [CrossRef]
48. Smith, E.L.; Abbott, A.P.; Ryder, K.S. Deep Eutectic Solvents (DESs) and their applications. *Chem. Rev.* **2014**, *114*, 11060–11082. [CrossRef] [PubMed]
49. Paiva, A.; Craveiro, R.; Aroso, I.; Martins, M.; Reis, R.L.; Duarte, A.R.C. Natural deep eutectic solvents—Solvents for the 21st century. *ACS Sustain. Chem. Eng.* **2014**, *2*, 1063–1071. [CrossRef]
50. Di Gioia, M.L.; Cassano, R.; Costanzo, P.; Herrera Cano, N.; Maiuolo, L.; Nardi, M.; Nicoletta, F.P.; Oliverio, M.; Procopio, A. Green Synthesis of Privileged Benzimidazole Scaffolds Using Active Deep Eutectic Solvent. *Molecules* **2019**, *24*, 2885. [CrossRef] [PubMed]
51. Bonacci, S.; Di Gioia, M.L.; Costanzo, P.; Maiuolo, L.; Tallarico, S.; Nardi, M. Natural Deep Eutectic Solvent as Extraction Media for the Main Phenolic Compounds from Olive Oil Processing Wastes. *Antioxidants* **2020**, *9*, 513. [CrossRef]
52. Leitner, W.; Poliakoff, M. Supercritical fluids in green chemistry. *Green Chem.* **2008**, *10*, 730.
53. Carlès, P. A brief review of the thermophysical properties of supercritical fluids. *J. Supercrit. Fluids* **2010**, *53*, 2–11. [CrossRef]
54. Lindström, U.M. Stereoselective Organic Reactions in Water. *Chem. Rev.* **2002**, *10*, 2751–2772. [CrossRef] [PubMed]
55. Procopio, A.; Gaspari, M.; Nardi, M.; Oliverio, M.; Tagarelli, A.; Sindona, G. Simple and efficient MW-assisted cleavage of acetals and ketals in pure water. *Tetrahedron Lett.* **2007**, *48*, 8623–8627. [CrossRef]
56. Procopio, A.; Gaspari, M.; Nardi, M.; Oliverio, M.; Rosati, O. Highly efficient and versatile chemoselective addition of amines to epoxides in water catalyzed by erbium(III) triflate. *Tetrahedron Lett.* **2008**, *49*, 2289–2293. [CrossRef]
57. Simon, M.O.; Li, C.J. Green chemistry oriented organic synthesis in water. *Chem. Soc. Rev.* **2012**, *41*, 1415–1427. [CrossRef] [PubMed]
58. Oliverio, M.; Costanzo, P.; Paonessa, R.; Nardi, M.; Procopio, A. Catalyst-free tosylation of lipophilic alcohols in water. *RSC Adv.* **2013**, *3*, 2548–2552. [CrossRef]
59. Nardi, M.; Herrera Cano, N.; Costanzo, P.; Oliverio, M.; Sindona, G.; Procopio, A. Aqueous MW eco-friendly protocol for amino group protection. *RSC Adv.* **2015**, *5*, 18751–18760. [CrossRef]
60. Nardi, M.; Di Gioia, M.L.; Costanzo, P.; De Nino, A.; Maiuolo, L.; Oliverio, M.; Olivito, F.; Procopio, A. Selective acetylation of small biomolecules and their derivatives catalyzed by Er(OTf)$_3$. *Catalysts* **2017**, *7*, 269. [CrossRef]
61. Nardi, M.; Costanzo, P.; De Nino, A.; Di Gioia, M.L.; Olivito, F.; Sindona, G.; Procopio, A. Water excellent solvent for the synthesis of bifunctionalized cyclopentenones from furfural. *Green Chem.* **2017**, *19*, 5403–5411. [CrossRef]
62. Olivito, F.; Costanzo, P.; Di Gioia, M.L.; Nardi, M.; Oliverio, M.; Procopio, A. Efficient synthesis of organic thioacetate in water. *Org. Biomol. Chem.* **2018**, *16*, 7753–7759. [CrossRef]
63. Procopio, A.; De Luca, G.; Nardi, M.; Oliverio, M.; Paonessa, R. General MW-assisted grafting of MCM-41: Study of the dependence on time dielectric heating and solvent. *Green Chem.* **2009**, *11*, 770–773. [CrossRef]
64. Estevão, M.S.; Afonso, C.A.M. Synthesis of trans-4,5-diaminocyclopent-2-enones from furfural catalyzed by Er(III) immobilized on silica. *Tetrahedron Lett.* **2017**, *58*, 302–304. [CrossRef]
65. Senthilkumar, S.; Maru, M.S.; Somani, R.S.; Bajaj, H.C.; Neogi, S. Unprecedented NH$_2$-MIL-101(Al)/*n*-Bu$_4$NBr system as solvent-free heterogeneous catalyst for efficient synthesis of cyclic carbonates via CO$_2$ cycloaddition. *Dalton Trans.* **2018**, *47*, 418–428. [CrossRef] [PubMed]
66. Mason, T.J. Sonochemistry: Current uses and future prospects in the chemical and processing industries. *Philos. Trans. R. Soc. Lond. A* **1999**, *357*, 355–369. [CrossRef]
67. Cravotto, G.; Cintas, P. The combined use of microwaves and ultrasound: Improved tools in process chemistry and organic synthesis. *Chem. Eur. J.* **2007**, *13*, 1902–1909. [CrossRef] [PubMed]
68. Je˘selnik, M.; Varma, R.S.; Polanca, S.; Kocevar, M. Catalyst-free reactions under solvent-free conditions: Microwave-assisted synthesis of heterocyclic hydrazones below the melting points of neat reactants. *Chem. Commun.* **2001**, *18*, 1716–1717. [CrossRef]

69. Kappe, O. Controlled microwave heating in modern organic synthesis. *Angew. Chem. Int. Ed.* **2004**, *43*, 6250–6284. [CrossRef]
70. Desai, K.R. *Green Chemistry Microwave Synthesis*, 1st ed.; Himalaya Publication House: New Delhi, India, 2005; p. 20.
71. Horikoshi, S.; Serpone, N. Role of microwaves in heterogeneous catalytic systems. *Catal. Sci. Technol.* **2014**, *4*, 1197–1210.
72. Procopio, A.; Dalpozzo, R.; De Nino, A.; Maiuolo, L.; Nardi, M.; Romeo, G. Mild and efficient method for the cleavage of benzylidene acetals by using erbium (III) triflate. *Org. Biomol. Chem.* **2005**, *3*, 4129–4133. [CrossRef]
73. Oliverio, M.; Costanzo, P.; Nardi, M.; Calandruccio, C.; Salerno, R.; Procopio, A. Tunable microwave-assisted method for the solvent-free and catalyst-free peracetylation of natural products. *Beilstein J. Org. Chem.* **2016**, *12*, 2222–2233. [CrossRef]
74. Maiuolo, L.; Merino, P.; Algieri, V.; Nardi, M.; Di Gioia, M.L.; Russo, B.; Delso, I.; Tallarida, M.A.; De Nino, A. Nitrones and nucleobase-containing spiro-isoxazolidines derived from isatin and indanone: Solvent-free microwave-assisted stereoselective synthesis and theoretical calculations. *RSC Adv.* **2017**, *7*, 48980–48988. [CrossRef]
75. Bortolini, O.; D'Agostino, M.; De Nino, A.; Maiuolo, L.; Nardi, M.; Sindona, G. Solvent-free, microwave assisted 1,3-cycloaddition of nitrones with vinyl nucleobases for the synthesis of N,O-nucleosides. *Tetrahedron* **2008**, *64*, 8078–8081. [CrossRef]
76. Procopio, A.; Gaspari, M.; Nardi, M.; Oliverio, M.; Romeo, R. MW-assisted Er(OTf)3 -catalyzed mild cleavage of isopropylidene acetals in Tricky substrates. *Tetrahedron Lett.* **2008**, *49*, 1961–1964. [CrossRef]
77. Nardi, M.; Bonacci, S.; De Luca, G.; Maiuolo, J.; Oliverio, M.; Sindona, G.; Procopio, A. Biomimetic synthesis and antioxidant evaluation of 3,4-DHPEA-EDA [2-(3,4-hydroxyphenyl) ethyl (3S,4E)-4-formyl-3-(2-oxoethyl)hex-4-enoate]. *Food Chem.* **2014**, *162*, 89–93. [CrossRef] [PubMed]
78. Oliverio, M.; Nardi, M.; Cariati, L.; Vitale, E.; Bonacci, S.; Procopio, A. "on Water" MW-Assisted Synthesis of Hydroxytyrosol Fatty Esters. *ACS Sustain. Chem. Eng.* **2016**, *4*, 661–665. [CrossRef]
79. Maiuolo, L.; De Nino, A.; Algieri, V.; Nardi, M. Microwave-assisted 1,3-dipolar cyclo-addition: Recent advances in synthesis of isoxazolidines. *Mini-Rev. Org. Chem.* **2017**, *14*, 136–142. [CrossRef]
80. Nardi, M.; Bonacci, S.; Cariati, L.; Costanzo, P.; Oliverio, M.; Sindona, G.; Procopio, A. Synthesis and antioxidant evaluation of lipophilic oleuropein aglycone derivatives. *Food Funct.* **2017**, *8*, 4684–4692. [CrossRef]
81. Costanzo, P.; Calandruccio, C.; Di Gioia, M.L.; Nardi, M.; Oliverio, M.; Procopio, A. First multicomponent reaction exploiting glycerol carbonate synthesis. *J. Clean. Prod.* **2018**, *202*, 504–509. [CrossRef]
82. Costanzo, P.; Bonacci, S.; Cariati, L.; Nardi, M.; Oliverio, M.; Procopio, A. Simple and efficient sustainable semi-synthesis of oleacein [2-(3,4-hydroxyphenyl) ethyl (3S,4E)-4-formyl-3-(2-oxoethyl)hex-4-enoate] as potential additive for edible oils. *Food Chem.* **2018**, *245*, 410–414. [CrossRef]
83. Paonessa, R.; Nardi, M.; Di Gioia, M.L.; Olivito, F.; Oliverio, M.; Procopio, A. Eco-friendly synthesis of lipophilic EGCG derivatives and antitumor and antioxidant evaluation. *Nat. Prod. Commun.* **2018**, *9*, 1117–1122. [CrossRef]
84. Li, J.-T.; Xing, C.-Y.; Li, T.-S. An efficient and environmentally friendly method for synthesis of arylmethylenemalononitrile catalyzed by Montmorillonite K10–ZnCl$_2$ under ultrasound irradiation. *J. Chem. Technol. Biotechnol.* **2004**, *79*, 1275–1278. [CrossRef]
85. Bhattacharyya, K.G.; Gupta, S.S. Adsorption of a few heavy metals on natural and modified kaolinite and montmorillonite: A review. *Adv. Colloid Interface Sci.* **2008**, *140*, 114–131. [CrossRef] [PubMed]
86. Kaur, N.; Kishore, D. Montmorillonite: An efficient, heterogeneous and green catalyst for organic synthesis. *J. Chem. Pharm. Res.* **2012**, *4*, 991–1015.
87. Kumar, B.S.; Dhakshinamoorthy, A.; Pitchumani, K. K10 montmorillonite clays as environmentally benign catalysts for organic reactions. *Catal. Sci. Technol.* **2014**, *4*, 2378–2396. [CrossRef]
88. Hechelski, M.; Ghinet, A.; Brice Louvel, B.; Dufrenoy, P.; Rigo, B.; Daïch, A.; Waterlot, C. From Conventional Lewis Acids to Heterogeneous Montmorillonite K10: Eco-Friendly Plant-Based Catalysts Used as Green Lewis Acids. *ChemSusChem* **2018**, *11*, 1249–1277. [CrossRef] [PubMed]
89. Rostamizadeh, S.; Amani, A.M.; Aryan, R.; Ghaieni, H.R.; Norouzi, L. Very fast and efficient synthesis of some novel substituted 2-arylbenzimidazoles in water using ZrOCl$_2$·nH$_2$O on montmorillonite K10 as catalyst. *Mon. Chem.* **2009**, *140*, 547–552. [CrossRef]

90. Hashemi, M.M.; Eftekhari-Sis, B.; Abdollahifar, A.; Khalili, B. ZrOCl2·8H2O on montmorillonite K10 accelerated conjugate addition of amines to α,β-unsaturated alkenes under solvent-free conditions. *Tetrahedron* **2006**, *62*, 672–677. [CrossRef]
91. Borah, S.J.; Das, D.K. Modified Montmorillonite: An Active Heterogeneous Catalyst for the Synthesis of Benzimidazoles. *J. Chem. Pharm. Res.* **2018**, *10*, 118–123.
92. Bonacci, S.; Nardi, M.; Costanzo, P.; De Nino, A.; Di Gioia, M.L.; Oliverio, M.; Procopio, A. Montmorillonite K10-Catalyzed Solvent-Free Conversion of Furfural into Cyclopentenones. *Catalysts* **2019**, *9*, 301. [CrossRef]
93. Hegedüs, A.; Hell, Z.; Potor, A. Zeolite-Catalyzed Environmentally Friendly Synthesis of Benzimidazole Derivatives. *Synth. Commun.* **2009**, *36*, 3625–3630. [CrossRef]
94. Khanday, W.A.; Tomar, R. Conversion of zeolite—A in to various ion-exchanged catalytic forms and their catalytic efficiency for the synthesis of benzimidazole. *Catal. Commun.* **2014**, *43*, 141–145. [CrossRef]
95. Saberi, A. Efficient synthesis of Benzimidazoles using zeolite, alumina and silica gel under microwave irradiation. *Iran. J. Sci. Technol.* **2015**, *39*, 7–10.
96. Nardi, M.; Cozza, A.; Maiuolo, L.; Oliverio, M.; Procopio, A. 1,5-Benzoheteroazepines through eco-friendly general condensation reactions. *Tetrahedron Lett.* **2011**, *52*, 4827–4834. [CrossRef]
97. Nardi, M.; Cozza, A.; De Nino, A.; Oliverio, M.; Procopio, A. One-pot synthesis of dibenzo[b,e][1,4]diazepin-1-ones. *Synthesis* **2012**, *44*, 800–804. [CrossRef]
98. Oliverio, M.; Costanzo, P.; Nardi, M.; Rivalta, I.; Procopio, A. Facile ecofriendly synthesis of monastrol and its structural isomers via biginelli reaction. *ACS Sustain. Chem. Eng.* **2014**, *2*, 1228–1233. [CrossRef]
99. Herrera Cano, N.; Uranga, J.G.; Nardi, M.; Procopio, A.; Wunderlin, D.A.; Santiago, A.N. Selective and eco-friendly procedures for the synthesis of benzimidazole derivatives. The role of the Er(OTf)$_3$ catalyst in the reaction selectivity. *Beilstein J. Org. Chem.* **2016**, *12*, 2410–2419. [CrossRef] [PubMed]
100. De Nino, A.; Maiuolo, L.; Nardi, M.; Pasceri, R.; Procopio, A.; Russo, B. Development of one-pot three component reaction for the synthesis of N′-aryl-N-cyanoformamidines, essential precursors of formamidine pesticides family. *Arab. J. Chem.* **2016**, *9*, 32–37. [CrossRef]

© 2020 by the authors. Licensee MDPI, Basel, Switzerland. This article is an open access article distributed under the terms and conditions of the Creative Commons Attribution (CC BY) license (http://creativecommons.org/licenses/by/4.0/).

Article

CoMn Catalysts Derived from Hydrotalcite-Like Precursors for Direct Conversion of Syngas to Fuel Range Hydrocarbons

Zahra Gholami *, Zdeněk Tišler, Romana Velvarská and Jaroslav Kocík

Unipetrol Centre of Research and Education, a.s, Areál Chempark 2838, Záluží 1, 436 70 Litvínov, Czech Republic; zdenek.tisler@unicre.cz (Z.T.); romana.velvarska@unicre.cz (R.V.); jaroslav.kocik@unicre.cz (J.K.)
* Correspondence: Zahra.gholami@unicre.cz; Tel.: +420-471-122-239

Received: 8 July 2020; Accepted: 19 July 2020; Published: 22 July 2020

Abstract: Two different groups of CoMn catalysts derived from hydrotalcite-like precursors were prepared through the co-precipitation method, and their performance in the direct production of gasoline and jet fuel range hydrocarbons through Fischer–Tropsch (FT) synthesis was evaluated in a batch autoclave reactor at 240 °C and 7 MPa and H_2/CO of 2. The physicochemical properties of the prepared catalysts were investigated and characterized using different characterization techniques. Catalyst performance was significantly affected by the catalyst preparation method. The crystalline phase of the catalyst prepared using KOH contained Co_3O_4 and some $Co_2MnO_{4.5}$ spinels, with a lower reducibility and catalytic activity than cobalt oxide. The available cobalt active sites are responsible for the chain growth, and the accessible acid sites are responsible for the cracking and isomerization. The catalysts prepared using $KOH + K_2CO_3$ mixture as a precipitant agent exhibited a high selectivity of 51–61% for gasoline (C_5–C_{10}) and 30–50% for jet fuel (C_8–C_{16}) range hydrocarbons compared with catalysts precipitated by KOH. The CoMn-HTC-III catalyst with the highest number of available acid sites showed the highest selectivity to C_5–C_{10} hydrocarbons, which demonstrates that a high Brønsted acidity leads to the high degree of cracking of FT products. The CO conversion did not significantly change, and it was around 35–39% for all catalysts. Owing to the poor activity in the water-gas shift reaction, CO_2 formation was less than 2% in all the catalysts.

Keywords: Fischer-Tropsch; jet fuel; gasoline; CoMn; hydrotalcite-like precursors

1. Introduction

The impending depletion of fossil fuel sources and the growing demand for energy resources because of increasing population and economic development have led to new approaches to the production of renewable liquid fuels. X-to-liquid technologies for converting different carbon-containing sources, such as natural gas (GTL), coal (CTL), biomass (BTL), and waste/oil residues (WTL), to liquid fuels have received special attention [1]. Typical transformation processes for the conversion of non-petroleum carbon resources into liquid fuel are shown in Figure 1. Carbonaceous resources are transformed into syngas (H_2 + CO) through reforming, gasification, or partial oxidation and then converted to a wide range of hydrocarbons. These hydrocarbons are refined to produce final products, including liquefied petroleum gas, gasoline, jet fuel, distillate, diesel, and wax [2]. Fischer–Tropsch (FT) is a well-known process for the catalytic conversion of syngas into higher hydrocarbons and oxygenates, which are finally upgraded to sulfur and aromatic free transportation fuels and chemicals [3–7]. The FT process plays an essential role in the production of sustainable and clean liquid fuels through CO hydrogenation. The product distribution of the traditional FT process follows the Anderson-Schulz-Flory (ASF) law and can be determined using the value of chain growth probability.

However, controlling product selectivity for the production of specific hydrocarbons, such as gasoline (C_5–C_{10}), diesel fuel (C_{11}–C_{22}), or jet fuel (C_8–C_{16}), is extremely challenging [8,9]. Several researchers have reported efficient methods for directly synthesizing gasoline, diesel, and jet fuel [8,10–13]. The direct conversion of syngas to liquid fuels through the FT process reduces the demand for refining units, such as hydrocrackers, by increasing the product selectivity of desired liquid hydrocarbons. Different parameters, such as the active phase of the catalyst and its chemical state, physicochemical properties, support, and promoter, considerably influence the catalytic activity and product selectivity in the FT reaction. Therefore, developing selective catalysts for the direct production of targeted products through the FT process has attracted considerable attention from scientists.

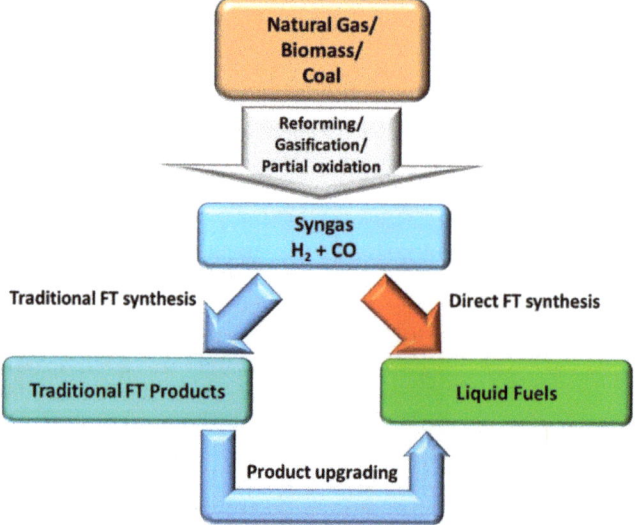

Figure 1. Transformation of carbonaceous resources (natural gas, biomass, and coal) into liquid fuels.

Cobalt-based catalysts with high selectivities to long-chain hydrocarbons, including diesel fuel and wax, are used more often than iron-based catalysts, which are suitable for gasoline production. Promoters can promote the reduction of cobalt oxide to its active metal phase and improve the catalyst's lifetime and mechanical stability by inhibiting carbon deposition on active Co^0 and decreasing Co sintering [14]. The 25% Co/Al_2O_3 catalysts promoted by Pt have been used as an active catalyst for the production of aviation fuel in a continuous stirred tank reactor (CSTR) at 1.8 MPa and 220 °C [15], and a CO conversion rate of approximately 30% and selectivity of 28%, 17%, and 40% to C_5–C_{11}, C_{12}–C_{18}, and C_{19+}, respectively, was observed. In another study, a Co/ZrO_2-SiO_2 catalyst prepared through the incipient wetness impregnation method and its catalytic performance in the direct synthesis of jet fuel from syngas in a slurry phase reactor was investigated by Li et al. [8]. A C_8–C_{16} selectivity of 29% was obtained at 1 MPa and 240 °C. However, the addition of a co-fed additive of syngas containing 1-decene and 1-tetradecene (1:1) resulted in a significant increase in selectivity. The addition of Mn to Co/TiO_2 catalysts increases C_{5+} selectivity (from 30.9% to 43.5%) and decreases methane selectivity (from 32.7% to 22.3%) [16]. Co–Mn interactions in catalysts prepared through homogeneous deposition precipitation (HDP) decrease the Co reducibility. The decrease in reducibility did not affect catalyst activity; rather, it improved selectivity to higher hydrocarbons (C_{5+}). The addition of Mn to Co catalysts prepared through an impregnation method did not cause changes in reduction temperature, probably because of the absence of interactions between Mn and Co in these catalysts [16].

Hydrotalcite-like compounds (HTCs) comprise atomically dispersed mixed metals and are excellent precursors and/or catalyst supports [17–19]. HTCs are brucite-like layer materials with a general formula of $(M^{2+}_{1-x}M^{3+}_x(OH)_2)^{x+}(A^{n-}_{x/n})^{x-}\cdot mH_2O$, where partial metal cation M^{2+}/M^{3+} replacement occurs and the excess of positive charge is counterbalanced by anions A^{n-} (such as CO_3^{2-}, SO_4^{2-}, NO_3^-, or other organic anions such as terephthalate) existing in interlayers, together with water molecules, and x = $M^{3+}/(M^{2+} + M^{3+})$, which generally ranges from 0.2 to 0.33, is the surface charge determined by the ratio of two metal cations; it can be changed for different applications [20,21]. HTCs can be synthesized as catalyst precursors with divalent metal cations (M^{2+}) (e.g., Mg^{2+}, Ni^{2+}, Co^{2+}, Cu^{2+}, or Zn^{2+}) and trivalent cations (M^{3+}) (e.g., Mn^{3+}, Al^{3+}, Fe^{3+}, Cr^{3+}, Rh^{3+}, Ru^{3+}, Ga^{3+} or In^{3+}). The calcination of HTCs leads to their transformation into well-dispersed mixed metal oxides (MMOs) with high surface areas, numerous Lewis base sites, and good thermal stability against sintering. These features are highly suitable for catalysis applications. The formation of O^{2-}–M^{n+} acid-base pairs is associated with the types of acidic-basic sites. The Lewis acidity is associated with the presence of low-coordination O^2 species that closely interact with M^{3+} cations, and Lewis basicity is due to M^{2+} cations [22,23]. The nature and strength of the acidic-basic sites of hydrotalcite-like compounds can be adjusted by adjusting the following parameters: (1) the nature of substituting cations in hydrotalcite structures; (2) characteristic M^{2+}/M^{3+} molar ratio; (3) nature of anions presented in interlayer regions; (4) thermal activation of layered materials; for example, a higher calcination temperature is favored for the creation of Lewis basic sites) [22,23].

The reduction of HTCs can be eminently suitable for the formation of highly dispersed and well-supported metallic particles [17–21,24–26]. Hydrotalcite-derived mixed oxides have been used as a catalyst in different catalytic reactions, such as steam reforming of ethanol [20,27,28], CO_2 hydrogenation to methanol [29,30], CO_2 reforming of methane for syngas production [31–33], formation of alcohols from syngas [19,34,35], and hydrocarbon production through FT reaction [36]. In the present study, CoMn catalysts derived from hydrotalcite-like precursors, with a Co/Mn molar ratio of 2, were synthesized using different preparation methods, and their catalytic activities in the FT reaction were evaluated. The physicochemical properties of the catalysts were characterized through thermogravimetric analysis (TG), inductively coupled plasma (ICP), scanning electron microscopy (SEM), hydrogen temperature-programmed reduction (H_2-TPR), ammonia temperature-programmed desorption (NH_3-TPD), and X-ray diffraction (XRD), and the effects of different preparation methods on the structures and catalytic performance of the catalysts were investigated.

2. Results and Discussion

2.1. Characterization of Catalysts

The inductively coupled plasma-optical emission spectrometry (ICP-OES) analysis was performed to determine the chemical composition of the prepared catalysts, and the results are listed in Table 1. The Co/Mn molar ratios calculated from the ICP analysis are close to the theoretical value (Co/Mn = 2), indicating the complete precipitation of metal ions. The thermogravimetric analysis (TGA) and derivative thermogravimetric (DTG) curves for the CoMn catalysts derived from hydrotalcite-like precursors (CoMn-HTC catalysts) (Figure 2) exhibited several phases of weight loss because of the thermal decomposition of catalysts, and the same behavior was observed in the DTG curves. The peaks at 100–260 °C were attributed to the evaporation of physically adsorbed water and interpolated water molecules associated with the dried samples. This result is in good agreement with the mass spectroscopy results, showing that water was released at this temperature range, and weight loss mainly occurred at this range. The weak peak observed at higher temperatures was caused by the elimination of hydroxyl and carbonate anions from the interlayer space, along with interlayer water. Weight loss in the catalysts precipitated with KOH (group 2) was lower than that in the catalysts prepared using a mixture of KOH + K_2CO_3 as a precipitating agent. Compared with the other catalysts in each group, the catalysts prepared with the addition of H_2O_2 had a lower weight loss than the other catalysts. The peaks for released CO_2 were observed at around 280 °C and 240 °C for catalysts I and II, and no apparent peaks were observed for the other catalysts. The amount of released CO_2 is much lower than that of released water. The low amount of released CO_2 may be due to the adsorption of CO_2 from air during the measurement.

Table 1. Inductively coupled plasma (ICP) analysis of the CoMn-HTC catalysts.

Samples	Concentration (wt.%)		Concentration (mol%)		Molar Ratio
	Co	Mn	Co	Mn	Co/Mn
CoMn-HTC-I.	41.3	14.7	0.70	0.27	2.62
CoMn-HTC-II.	41.0	15.2	0.70	0.28	2.51
CoMn-HTC-III.	42.3	19.5	0.72	0.35	2.02
CoMn-HTC-IV.	44.1	18.6	0.75	0.34	2.21
CoMn-HTC-V.	41.2	16.6	0.70	0.30	2.31
CoMn-HTC-VI.	43.3	18.6	0.73	0.34	2.17

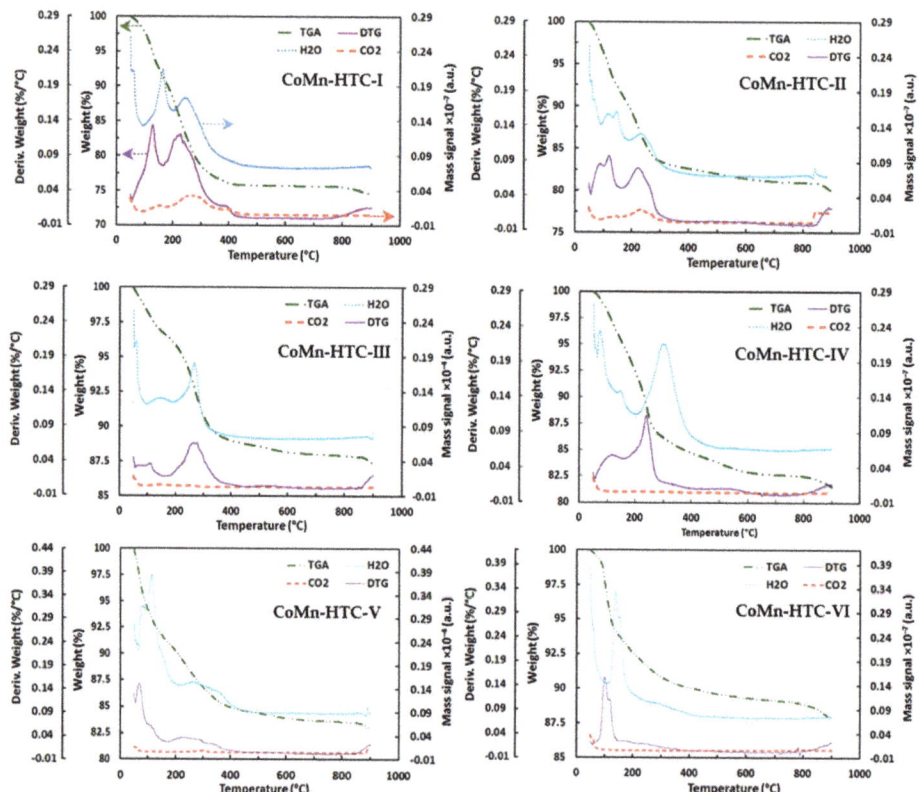

Figure 2. Thermogravimetric analysis and mass spectrometry detector (TGA-MS) analysis of the CoMn-HTC catalysts.

The reducibility of CoMn-HTC catalysts was analyzed through hydrogen temperature-programmed reduction (H_2-TPR) analysis (Figure 3). The TPR profiles revealed two regions of low-temperature range at 180–450 °C and high-temperature range at 450–730 °C. The first weak peak in the range of 199–237 °C was ascribed to the partial reduction of easily reducible species of CoMn composite oxide:

$$Mn_xCo_yO_z(s) + H_2(g) \rightarrow Mn_3O_4\text{-}Co_3O_4(s) + H_2O(g) \tag{1}$$

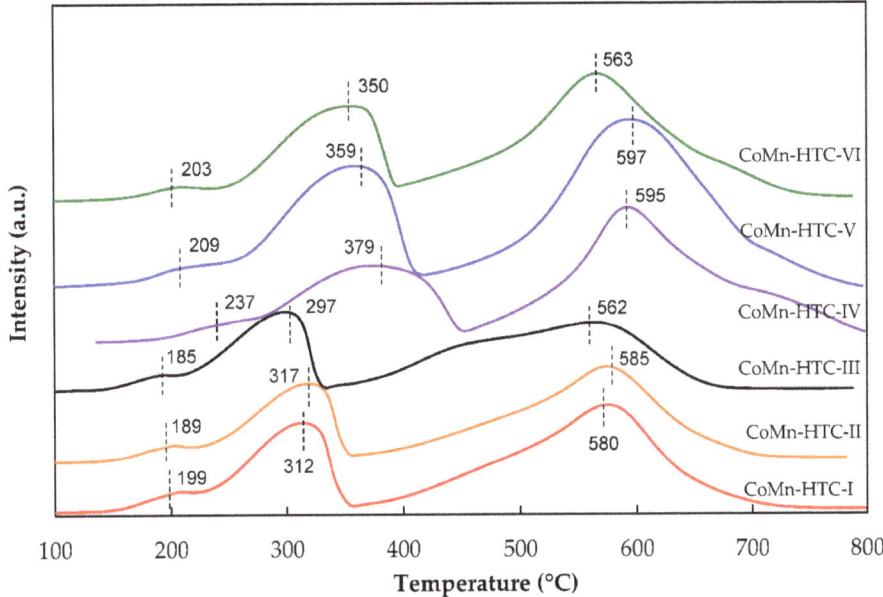

Figure 3. Hydrogen temperature-programmed reduction (H$_2$-TPR) profiles of the CoMn-HTC catalysts.

The second peak at 300–380 °C was attributed to the further reduction of Mn$_3$O$_4$ to MnO and Co$_3$O$_4$ to CoO:

$$Mn_3O_4\text{-}Co_3O_4(s) + H_2(g) \rightarrow MnO\text{-}CoO(s) + H_2O(g) \quad (2)$$

The peaks at high temperature (>450 °C) were ascribed to the reduction of CoO to Co and Mn$_3$O$_4$ to MnO:

$$MnO\text{-}CoO(s) + H_2(g) \rightarrow MnO(s) + Co(s) + H_2O(g) \quad (3)$$

Compared with the TPR peaks of the catalysts prepared by KOH as the precipitating agent, the TPR peaks for the catalysts prepared by the mixture of KOH + K$_2$CO$_3$ as the precipitating agent shifted to lower temperatures. Precipitation in the presence of air did not cause a considerable change in catalyst II compared with catalyst I, while the peaks of catalysts V shifted to lower temperatures than those of catalyst IV. In CoMn-HTC III (from group 1) and CoMn-HTC VI (from group 2) catalysts (catalysts prepared with the addition of H$_2$O$_2$), all peaks shifted to lower temperatures compared with the catalysts of each group prepared without the addition of air or H$_2$O$_2$. The different TPR profiles of CoMn-HTC catalysts could be due to the different Mn and Co species interactions, which are related to the preparation methods. The CoMn-HTC III catalyst had the lowest reduction temperatures. Thus, the preparation of catalysts precipitated by the mixture of KOH + K$_2$CO$_3$ in the presence of H$_2$O$_2$ enhanced the catalyst reducibility. Jung et al. [37] studied the effect of precipitants on nickel-based catalysts prepared through the co-precipitation method, and it was found that the catalysts prepared with K$_2$CO$_3$ as the precipitating agent had lower reduction temperatures and a higher hydrogen uptake than those precipitated with KOH. The catalyst prepared with K$_2$CO$_3$ also exhibited high pore volume and good catalytic activity for methane steam reforming [37].

The acidic properties of the prepared catalysts were determined through ammonia temperature-programmed desorption (NH$_3$-TPD) analysis (Figure 4). The peaks at low temperatures, below 200 °C, were attributed to weak acid sites or physically adsorbed ammonia, and the peaks at higher temperatures in the range of 200–400 °C, were associated with the medium interaction between Brønsted acid sites and NH$_3$. The peaks at temperatures above 400 °C belong to strong acid sites [38–40]. Table 2 shows the concentration of both weak and medium acid sites (mmol NH_3/g_{cat}) of the catalysts. The acid sites (weak and medium acid sites) decreased, and the peaks shifted to higher temperatures in the catalysts prepared with KOH. Medium acid sites increased in both catalyst groups (precipitated by a mixture of KOH + K_2CO_3, and KOH) when they were precipitated by the addition of H_2O_2 to enhance the oxidation of metals.

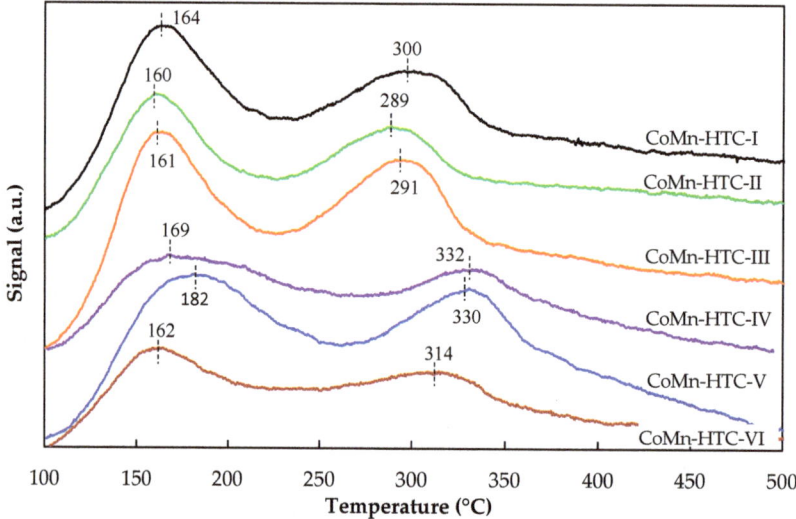

Figure 4. Ammonia temperature-programmed desorption (NH_3-TPD) profiles of the CoMn-HTC catalysts.

Table 2. Surface acidity of CoMn catalysts measured by ammonia temperature-programmed desorption (NH_3-TPD).

Catalyst	Acidic Site (mmol NH_3/g)		
	First Peak	Second Peak	Total
CoMn-HTC-I	0.172	0.276	0.448
CoMn-HTC-II	0.147	0.201	0.348
CoMn-HTC-III	0.184	0.353	0.537
Co-Mn-HTC-IV	0.212	0.104	0.316
CoMn-HTC-V	0.208	0.196	0.404
CoMn-HTC-VI	0.073	0.151	0.224

The higher number of acid sites could be attributed to the porous structure of the catalyst. The surface acidity of porous $Mn_2Co_1O_x$ catalysts prepared by the combustion (CB) and co-precipitation (CP) methods was studied by Qiao et al. [41]. They found that the ammonia desorption peaks shifted to a lower temperature region for the catalyst prepared by the co-precipitation method. The desorption of NH_4^+ ions bonded to Brønsted acid sites are easier at lower temperatures. For the catalyst prepared by the combustion method, with larger specific surface area and porous structures than those prepared by the co-precipitation method, the peaks shifted to slightly higher temperatures, indicating the presence of abundant Lewis acid sites; the stronger acid strength could be due to the stronger interaction between the cobalt oxide and manganese oxide species [41]. Since the acid sites enhance the cracking and isomerization of heavier hydrocarbons [12], the catalysts with higher acid sites are expected to have better performance for the production of lighter hydrocarbons. Liu et al. [42] reported that the

H$_2$O$_2$-modified catalysts have a larger number of surface acid sites, especially Brønsted acid sites. They reported that the NH$_3$-TPD peaks at low temperatures (<250 °C) are due to the desorption of physisorbed ammonia and partial ionic NH$_4^+$ bound to weak Brønsted acid sites, and the peaks at higher temperatures (>250 °C) belong to the Lewis acid sites and ionic NH$_4^+$ bound to Brønsted acid sites. Brønsted acid sites are ascribed to surface protons, whereas Lewis acid sites, which are stronger than Brønsted sites, are attributed to the Co-O-Mn species located within a CoO structure and containing Mn^{3+} cations mainly in octahedral sites.

The X-ray diffraction (XRD) patterns of the prepared catalysts are shown in Figure 5. The XRD patterns of dried catalysts (Figure 5a) showed that the CoMn-HTC catalysts had crystallized hydrotalcite structure forms, with characteristic peaks at 11.9°, 23.5°, 34.3°, 39°, 47.5°, 60.8°, and 62.9°, which correspond to the (0 0 3), (0 0 6), (0 0 9), (0 1 5), (0 1 8), (1 1 0), and (1 1 3) planes (JCPDS #70−2151) [43–45].

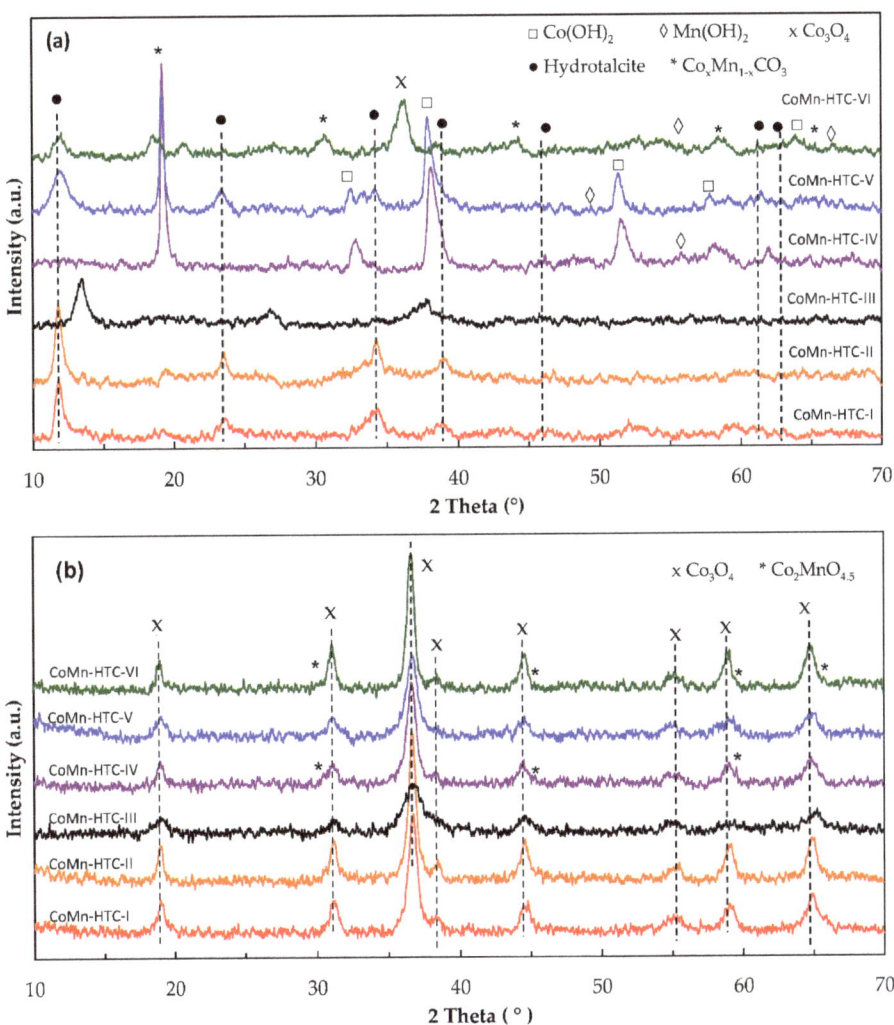

Figure 5. X-ray diffraction (XRD) patterns of the (a) dried and (b) calcined CoMn-HTC catalysts.

The CoMn-HTC-IV, CoMn-HTC-V, and CoMn-HTC-VI catalysts had no clear layered structure and did not show apparent peaks belonging to the hydrotalcite structure. Most peaks were attributed to CoMn composite carbonate ($Co_xMn_{1-x}CO_3$) in the catalysts prepared with KOH. The peaks of the hydrotalcite structure in the CoMn-HTC-III catalyst shifted to the right (mainly on the characteristic basal plane 0 0 3), and this can be attributed to the presence of anions in the interlayer regions. The difference in anions (carbonate or hydroxyl) occurred during the catalyst synthesis process, and H_2O_2 addition or air bubbling during the precipitation process affected the competition among anions occupying the interlayer region. The small peaks at 60.8° and 62.9° were attributed to the interlayers of carbonate and nitrate anions [46]. The presence of the peaks belonging to $Co(OH)_2$ and $Mn(OH)_2$ could be due to incomplete moisture removal during the drying process. After calcination at 300 °C in air, no peaks associated with the hydrotalcite phase were observed in the XRD patterns (Figure 5b), indicating the destruction of the layered structure and the transformation of the hydrotalcite phase into Co_3O_4 (JCPDS #42-1467) [21,46–48] in all prepared catalysts. The crystalline phase of the catalysts prepared by KOH contained some spinel $Co_2MnO_{4.5}$, with higher reduction temperatures and lower catalytic activities.

The scanning electron microscope (SEM) images of the CoMn-HTC catalysts after drying are shown in Figure 6. The effects of catalyst preparation methods on the catalyst morphology were investigated. The prepared CoMn hydrotalcites were formed by the accumulation of aggregated nanoparticles. The CoMn-HTC-III catalyst showed a more obvious plate-like layered structure than the other catalysts. This result showed that using a mixture of KOH + K_2CO_3 and the addition of H_2O_2 have a positive effect on the preparation of hydrotalcite structured catalysts. In addition, the CoMn-HTC-III catalyst had the lowest reduction temperature possible because of its better layered structure. The distributions of metals were examined by performing the energy-dispersive X-ray spectroscopy (EDX) mapping of the catalysts (Figure 7). Mn and Co were homogeneously distributed over the entirety of the catalyst particles. The dispersion of cobalt and manganese particles could be promoted by the addition of hydrogen peroxide during the catalyst preparation and enhances crystal grain growth. Cui et al. [49] and Liu et al. [42] also reported that the addition of H_2O_2 could promote the dispersion of metal particles and improve the crystal grain growth, and this consequently resulted in the formation of mixed oxides with good thermal stability.

Figure 6. Scanning electron microscope (SEM) images of the CoMn-HTC catalysts.

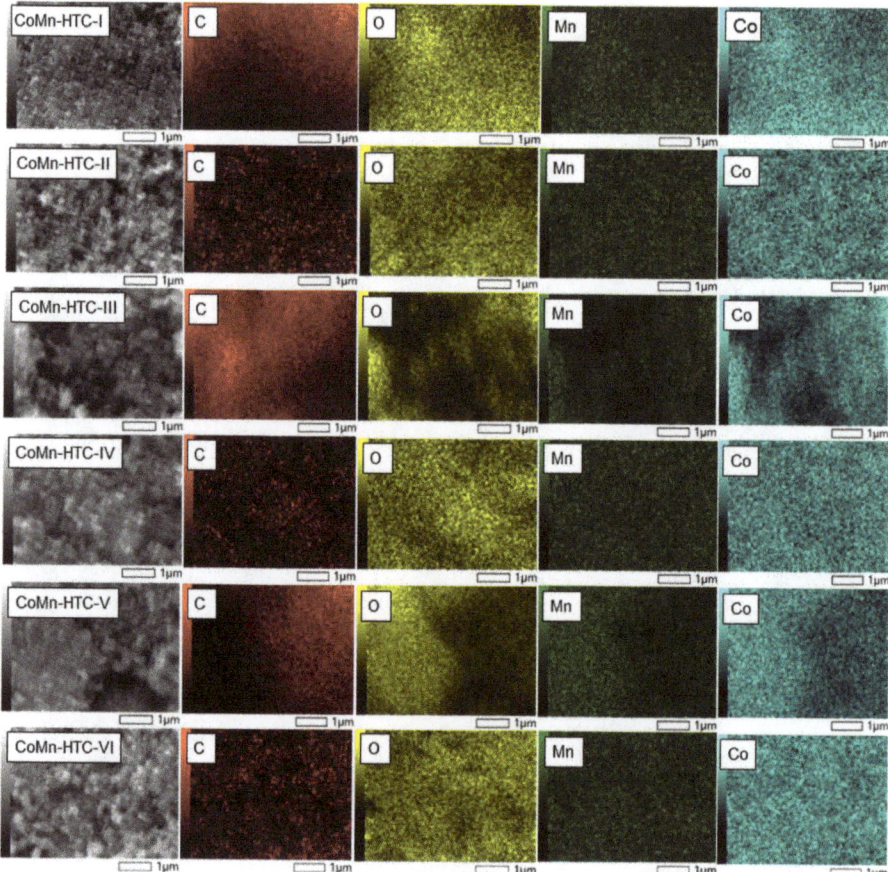

Figure 7. Energy-dispersive X-ray spectroscopy (EDX) mapping of the CoMn-HTC catalysts.

2.2. Catalytic Evaluation

Generally, FT synthesis products consist of light gaseous hydrocarbons (C_1–C_4), liquid fuels (C_5–C_{22}), and waxes (C_{23+}). The catalytic performance of calcined CoMn-HTC catalysts in FT reaction at H_2/CO = 2, with a reaction temperature of 240 °C, pressure of 7 MPa, and reaction time of 6 h are shown in Figure 8. The product distribution over two different groups of catalysts, (I, II, III) and (IV, V, VI), were different. The catalysts I, II, and III, which were prepared by a mixture of KOH + K_2CO_3 as the precipitating agent, showed a higher potential in the production of liquid fuels including gasoline (C_5–C_{10}) (Figure 8a) and jet fuel range hydrocarbons (C_8–C_{16}) (Figure 8b), whereas the other catalysts (IV, V, VI) showed a higher potential in the formation of heavier hydrocarbons.

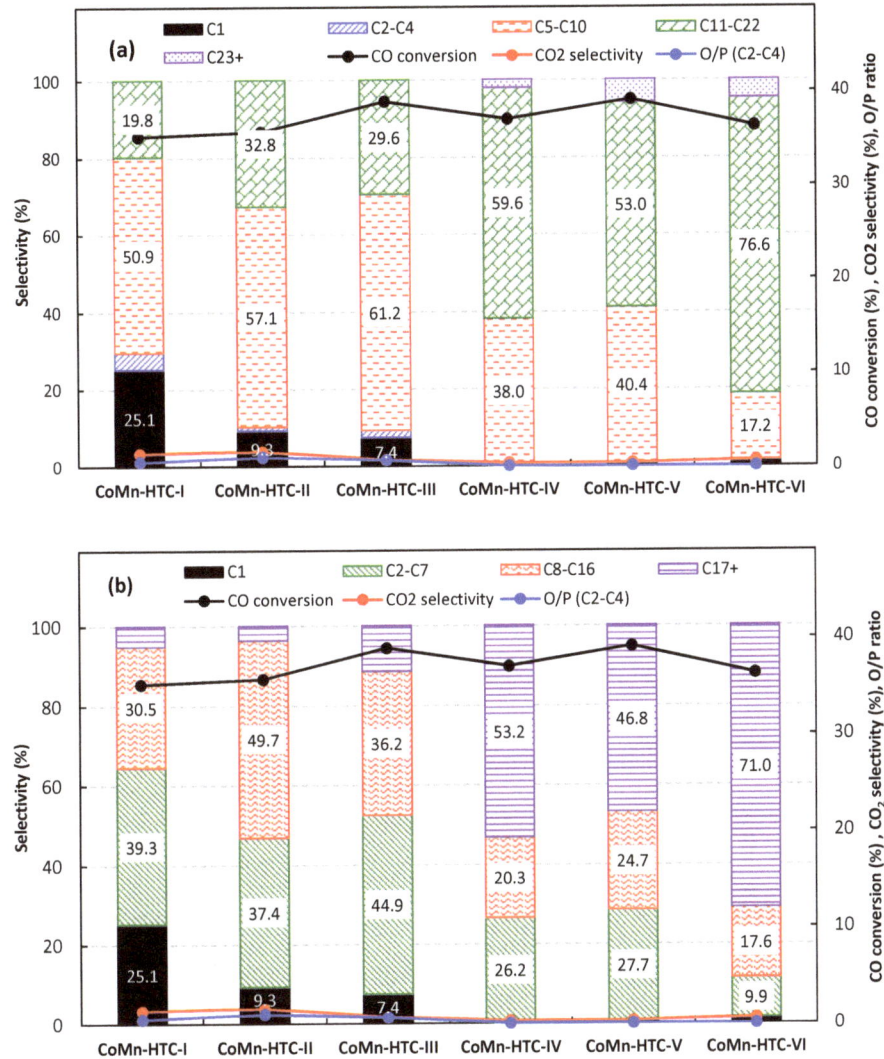

Figure 8. Product distribution in the Fischer–Tropsch (FT) reaction over CoMn-HTC catalysts: (a) gasoline range hydrocarbons (C_5–C_{10}), and (b) jet fuel range hydrocarbons (C_8–C_{16}). Reaction conditions: H_2/CO = 2, 240 °C, 7 MPa, 6 h.

The CoMn-HTC-I catalyst exhibited 48.6% selectivity to C_5–C_{10} and 30.5% to jet fuel. The CoMn-HTC-III catalyst had the highest selectivity of 61.2% to gasoline range hydrocarbons with the methane selectivity of 7.4%. The simultaneous addition of H_2O_2 during the precipitation process also leads to an increase in the CO conversion, from 35.2% to 38.9%. The CoMn-HTC-II catalyst had the highest selectivity (49.7%) for jet fuel range hydrocarbons. The product distributions on catalysts IV, V, and VI revealed that the methane selectivity decreased to less than 1%, and selectivities to C_5–C_{10} and C_8–C_{16} also decreased to 17–40% and 17–25%, respectively. The selectivity to C_{17+} in these catalysts was considerably higher than those in catalysts of group I, which were prepared by a mixture of KOH + K_2CO_3 as the precipitating agent. Very small amounts of olefins were detected over the CoMn-HTC catalysts at the given reaction conditions. The CO conversion did not change considerably over different catalysts, and it was in the range of 35–39% for all catalysts. The CO_2 formation rates in all catalysts were less than 2%, implying that the CoMn-HTC catalysts had extremely poor activity for the water–gas shift reaction (WGS: $CO + H_2O \rightarrow CO_2 + H_2$).

The reduction temperature of easily reducible species of CoMn composite oxide decreased in both air-bubbled and H_2O_2 added catalysts, and this decrease resulted in a slight increase in catalytic activity because of the formation of additional Mn^{3+} and more CoMn species, which are easy to reduce. In addition to the reducibility of the catalyst, accessible catalyst acid sites can greatly affect the product distribution. The in situ cracking of the FT products may have been affected by the surface acidity of the catalysts. In general, Brønsted acidity leads to the cracking/isomerization of FT waxes [12]. The integrated synthesis of gasoline, jet fuel, and diesel range hydrocarbons using cobalt catalysts supported on mesoporous Y-type zeolites were studied by Li et al. [12]; they found that the porosity and acidic properties of zeolites play an important role in product distribution, mainly affecting the chain growth and cracking of heavier hydrocarbons. The NH_3-TPD analysis revealed that the CoMn-HTC-III catalyst had higher Brønsted acid sites, which promoted the cracking of heavy hydrocarbons and the formation of C_5–C_{10} fraction. Catalysts I, II, and III, owing to the accessibility of their Brønsted acid sites and higher possibility for the cracking of heavy hydrocarbons, showed high selectivity to the gasoline range hydrocarbons (C_5–C_{10}) and jet fuel range hydrocarbons (C_8–C_{16}). However, excessive Brønsted acidity on the catalysts may have resulted in low catalyst stability and overcracking of heavy hydrocarbons and a subsequent increase in the fraction of lighter hydrocarbons and methane. Catalytic activity is strongly related to the types of precursor and precipitant agent. As it can be seen in Figure 8, a very small amount of olefinic products was produced during the reaction, the highest O/P ratio of C_2-C_4 hydrocarbon range of 1 was observed for catalyst II, and the highest O/P ratio in the whole range of produced hydrocarbon was less than 0.05. Oxygenated products were not detected in the products of the reaction for all catalysts.

According to Cai et al. [50], the activity of the copper manganese oxides catalysts prepared using sodium carbonate (Na_2CO_3) as the precipitating agent was higher than the catalyst precipitated by sodium hydroxide (NaOH) for the CO oxidation reaction. The crystalline phase of catalysts prepared using strong electrolyte (OH^-) was found to be mainly spinel $Cu_{1.5}Mn_{1.5}O_4$; catalysts prepared using weak electrolyte (CO_3^{2-}) mainly consisted of $MnCO_3$, Mn_2O_3, and CuO. The catalytic activities of the multiphase catalysts (containing CuO and Mn_2O_3) were several times higher than those of single-phase oxides, especially at low temperature ranges [50].

It is worth mentioning that the available cobalt active sites of the catalyst are responsible for the chain growth, and the accessible acid sites are responsible for the cracking and isomerization. In this study, it was observed that the first group of catalysts (I, II, and III), with more available active sites at lower temperatures and more available acid sites (mainly medium strength acid sites), showed higher selectivity to C_5–C_{10} and C_8–C_{16} hydrocarbons than the second group of catalysts. Owing to their lower acidic properties and the inferior cracking rates of the heavy hydrocarbons, catalysts IV, V, and VI revealed higher selectivity for the production of hydrocarbons heavier than the gasoline or jet fuel range of hydrocarbons, and they also showed a considerably lower selectivity to methane.

3. Materials and Methods

3.1. Catalyst Preparation

CoMn-HTC catalysts (Co/Mn molar ration = 2) with layered structures were synthesized by the co-precipitation method. Solution A was prepared by mixing cobalt nitrate ($Co(NO_3)_2 \cdot 6H_2O$), manganese nitrates ($Mn(NO_3)_2 \cdot 4H_2O$), and ammonium fluoride (NH_4F) in a 2000 mL beaker at room temperature. The mixture was continuously stirred for 60 min. Two different basic solutions (solution B) containing (1) (KOH (2 mol/L) + K_2CO_3 (0.2 mol/L)) and (2) (KOH (2 mol/L)), were used as precipitating agents. Solution B was added dropwise to solution A with vigorous stirring, at pH = 9 and room temperature. Details of the catalyst preparation conditions are provided in Table 3.

Table 3. Catalyst preparation conditions.

Catalyst	Precipitating Agent
CoMn-HTC-I (Group 1)	$KOH + K_2CO_3$
CoMn-HTC-II (Group 1)	$KOH + K_2CO_3$ + air bubbling
CoMn-HTC-III (Group 1)	$KOH + K_2CO_3 + H_2O_2$
CoMn-HTC-IV (Group 2)	KOH
CoMn-HTC-V (Group 2)	KOH + air bubbling
CoMn-HTC-VI (Group 2)	$KOH + H_2O_2$

Catalysts I, II, and III were precipitated as hydrotalcite precursors with a $KOH + K_2CO_3$ solution, and catalysts IV, V, and VI were precipitated by KOH. In catalysts II and V, air was bubbled through precipitation for the oxidation of Mn^{2+} to Mn^{3+}. Catalysts III and VI were precipitated by the simultaneous addition of 100 mL of H_2O_2 (30 vol.%). The resulting solution was aged for 24 h at room temperature. The solid product was filtered and washed with deionized water to a neutral pH, then dried at room temperature. The dried products were then calcined at 300 °C for 5 h in air (3 °C/min). The obtained catalysts were denoted as CoMn-HTC I to VI.

3.2. Catalyst Characterizations

The bulk metal contents in the prepared catalysts were determined using inductively coupled plasma-optical emission spectrometry (ICP-OES; Agilent 725/Agilent Technologies Inc., Santa Clara, CA, USA). Before analysis, approximately 0.5 g of catalyst was dissolved in 10 mL aqueous solution of H_2SO_4 (1:1) and heated. Then, the solution was cooled down and diluted with demineralized water and heated to 100 °C for 2 min. The obtained solution was then used for ICP analysis.

Thermogravimetric analysis (TGA) was carried out using the TGA Discovery series (TA Instruments, Lukens Drive, NW, USA). Approximately 20 mg of the catalyst was placed in an open aluminum crucible and heated from 50 °C to 900 °C at 10 °C/min in a nitrogen atmosphere (20 mL/min). The fragments were detected using an OmniStar GSD320 quadrupole mass detector (Pfeiffer Vacuum Austria GmbH, Vienna, Austria).

The surface morphology of the prepared catalysts was studied using a scanning electron microscope (SEM) (JEOL JSM-IT500HR; JEOL Ltd., Tokyo, Japan) accessorized with energy-dispersive X-ray spectroscopy (EDX) for elemental analysis and map analysis. Representative backscattered electron or secondary electron images of microstructures were taken in high vacuum mode, using an accelerating voltage of 15 kV.

The X-ray diffraction (XRD) patterns of the prepared catalysts were measured using a D8 Advance ECO (Bruker AXC GmbH, Karlsruhe, Germany) with CuKα radiation (λ = 1.5406 Å). The step time was 0.5 s, and the step size was 0.02° in a 2θ angle ranging from 10° to 70°. The diffractograms were evaluated using the Diffrac.Eva software with the Powder Diffraction File database (PDF 4+ 2018, International Centre for Diffraction Data).

Hydrogen temperature-programmed reduction (H_2-TPR) and ammonia temperature-programmed desorption (NH_3-TPD) analysis were performed using an Autochem 2950 HP (Micromeritics Instrument Corporation, Norcross, GA, USA). In H_2-TPR analysis, 50 mg of catalyst was placed in the quartz tube. The catalyst sample was pretreated under argon flow at 450 °C (10 °C/min) for 30 min to remove traces of water and impurities from the catalyst pores, and it was then cooled to 40 °C. H_2-TPR was performed using 10% H_2/Ar with a flow rate of 30 mL/min and heating from 40 °C to 800 °C (10 °C/min). In NH_3-TPD analysis, approximately 0.1 g of catalyst was pretreated with He (25 mL/min) at 450 °C (10 °C/min) for 1 h. After cooling to 80 °C, ammonia adsorption was carried out, and the catalyst was saturated with 10%NH_3/He (25 mL/min) at 80 °C for 1 h. The physically bound molecules of ammonia were removed by purging with He (25 mL/min) at 100 °C for 1 h. Finally, NH_3 desorption was performed by increasing the temperature from 100 °C to 800 °C, at a heating rate of 10 °C/min under He, with a flow rate of 25 mL/min. A thermal conductivity detector (TCD) was used to detect desorbed NH_3 in the outlet gas.

3.3. Catalytic Evaluation

The FT test was carried out in a 1 L stainless steel autoclave batch reactor (Parr instruments). In a typical experiment, 500 mg of catalyst and 50 mL of cyclohexane as a solvent were added to the reactor vessel. For the in-situ reduction of catalyst, the reactor was sealed and purged with N_2 five times and with H_2 three times. The reactor was then heated to 300 °C at a ramping rate of 3 °C/min, then pressurized with H_2 to 5 MPa for 5 h. After reduction, the reactor system was cooled to room temperature and purged three times with premixed syngas (H_2/CO = 2). The reactor temperature was raised to 240 °C at a ramping rate of 3 °C/min, then pressurized to 7 MPa to conduct the reaction in batch mode under a constant stirring speed of 800 rpm to eliminate the diffusion control region. A constant temperature was maintained during the reaction for 6 h. The conversion rate was measured according to the decrease in pressure during the reaction. After the reaction was terminated, the products were analyzed using different chromatographic procedures. The resultant gas sample was transferred to a gas bag and analyzed with a gas chromatograph, Agilent 7890A, with three parallel channels which collect data at the same time. The channels are equipped with two thermal conductivity detectors (TCD), CO, H_2, N_2, and CO_2 gases, and a flame ionization detector (FID) for the detection of hydrocarbons. The liquid samples were analyzed, without prior preparation steps, on chromatograph Agilent 7890A with FID detector using non-polar column HP PONA.

4. Conclusions

The effects of catalyst preparation methods on the physicochemical properties of CoMn-HTC catalysts derived from hydrotalcite-like precursors were investigated. The characterization results showed that the catalysts precipitated with KOH as a strong electrolyte (OH^-) had higher reduction temperatures and lower reducibility, reduced and readily reducible species, and less accessible acid sites on the surface than the catalysts prepared with the KOH + K_2CO_3 mixture as precipitating agent and thus had better catalyst reducibility and more accessible medium and strong acid sites. The catalytic performance of the prepared CoMn catalysts was evaluated for FT synthesis at 240 °C, 7 MPa, and H_2/CO = 2 in a batch autoclave reactor. The highest selectivity (61%) was obtained after using the catalyst prepared with KOH + K_2CO_3 mixture and the addition of H_2O_2 (CoMn-HTC-III). The selectivity to gasoline range hydrocarbons decreased from 61% to 36% after the precipitating agent was changed from mixed solution to KOH only. The CO conversion did not greatly change by variations in catalyst preparation methods and remained in the range of 35–39% in all catalysts. CO_2 formation was less than 2% regardless of the catalyst used, indicating the extremely poor and negligible activity of the catalysts for the water-gas shift reaction.

Author Contributions: Conceptualization and experimental work designed and supported by Z.G. and Z.T.; the characterizations, catalyst evaluation, and analysis were done by Z.G., Z.T., R.V., and J.K.; the manuscript was written and amended by Z.G. and Z.T. All authors have read and agreed to the published version of the manuscript.

Funding: This publication is a result of the project CATAMARAN, Reg. No. CZ.02.1.01/0.0/0.0/16_013/0001801, which has been co-financed by the European Union from the European Regional Development Fund through the operational program, Research, Development, and Education. This project has also been financially supported by the Ministry of Industry and Trade of the Czech Republic, which has been providing institutional support for the long-term conceptual development of research organization. The project CATAMARAN has been integrated into the National Sustainability Programme I of the Ministry of Education, Youth and Sports of the Czech Republic (MEYS), through the project, Development of the UniCRE Centre (LO1606). The result was achieved using the infrastructure of the project Efficient Use of Energy Resources Using Catalytic Processes (LM2015039), which has been financially supported by MEYS within the targeted support of large infrastructures.

Conflicts of Interest: The authors declare no conflict of interest. The funders had no role in the design of the study; in the collection, analyses, or interpretation of data; in the writing of the manuscript, or in the decision to publish the results.

References

1. Vosoughi, V.; Badoga, S.; Dalai, A.K.; Abatzoglou, N. Modification of mesoporous alumina as a support for cobalt-based catalyst in Fischer-Tropsch synthesis. *Fuel Process. Technol.* **2017**, *162*, 55–65. [CrossRef]
2. Pratt, J.W. A Fischer-Tropsch Synthesis Reactor Model Framework for Liquid Biofuels Production. SANDIA Report: SAND2012-7848, 2012. Available online: https://pdfs.semanticscholar.org/6093/1b3ffd88e3183156e132b694c50dc179833a.pdf (accessed on 15 May 2020).
3. Okoye-Chine, C.G.; Moyo, M.; Liu, X.; Hildebrandt, D. A critical review of the impact of water on cobalt-based catalysts in Fischer-Tropsch synthesis. *Fuel Process. Technol.* **2019**, *192*, 105–129. [CrossRef]
4. Vosoughi, V.; Badoga, S.; Dalai, A.K.; Abatzoglou, N. Effect of pretreatment on physicochemical properties and performance of multiwalled carbon nanotube supported cobalt catalyst for Fischer–Tropsch synthesis. *Ind. Eng. Chem. Res.* **2016**, *55*, 6049–6059. [CrossRef]
5. Odunsi, A.O.; O'Donovan, T.S.; Reay, D.A. Dynamic modeling of fixed-bed Fischer-Tropsch reactors with phase change material diluents. *Chem. Eng. Technol.* **2016**, *39*, 2066–2076. [CrossRef]
6. Gholami, Z.; Tišler, Z.; Rubáš, V. Recent advances in Fischer-Tropsch synthesis using cobalt-based catalysts: A review on supports, promoters, and reactors. *Catal. Rev.* **2020**, 1–84. [CrossRef]
7. Gholami, Z.; Zabidi, N.A.M.; Gholami, F.; Ayodele, O.B.; Vakili, M. The influence of catalyst factors for sustainable production of hydrocarbons via Fischer-Tropsch synthesis. *Rev. Chem. Eng.* **2017**, *33*, 337–358. [CrossRef]
8. Li, J.; Yang, G.; Yoneyama, Y.; Vitidsant, T.; Tsubaki, N. Jet fuel synthesis via Fischer–Tropsch synthesis with varied 1-olefins as additives using Co/ZrO_2–SiO_2 bimodal catalyst. *Fuel* **2016**, *171*, 159–166. [CrossRef]
9. Zhou, W.; Cheng, K.; Kang, J.; Zhou, C.; Subramanian, V.; Zhang, Q.; Wang, Y. New horizon in C1 chemistry: Breaking the selectivity limitation in transformation of syngas and hydrogenation of CO_2 into hydrocarbon chemicals and fuels. *Chem. Soc. Rev.* **2019**, *48*, 3193–3228. [CrossRef]
10. Sartipi, S.; van Dijk, J.E.; Gascon, J.; Kapteijn, F. Toward bifunctional catalysts for the direct conversion of syngas to gasoline range hydrocarbons: H-ZSM-5 coated Co versus H-ZSM-5 supported Co. *Appl. Catal. A Gen.* **2013**, *456*, 11–22. [CrossRef]
11. Shi, D.; Faria, J.; Pham, T.N.; Resasco, D.E. Enhanced activity and selectivity of Fischer–Tropsch synthesis catalysts in water/oil emulsions. *ACS Catal.* **2014**, *4*, 1944–1952. [CrossRef]
12. Li, J.; He, Y.; Tan, L.; Zhang, P.; Peng, X.; Oruganti, A.; Yang, G.; Abe, H.; Wang, Y.; Tsubaki, N. Integrated tuneable synthesis of liquid fuels via Fischer–Tropsch technology. *Nat. Catal.* **2018**, *1*, 787–793. [CrossRef]
13. Sartipi, S.; Parashar, K.; Makkee, M.; Gascon, J.; Kapteijn, F. Breaking the Fischer–Tropsch synthesis selectivity: Direct conversion of syngas to gasoline over hierarchical Co/H-ZSM-5 catalysts. *Catal. Sci. Technol.* **2013**, *3*, 572–575. [CrossRef]
14. Choudhury, H.A.; Moholkar, V.S. Synthesis of liquid hydrocarbons by Fischer–Tropsch process using industrial iron catalyst. *Int. J. Innovat. Res. Sci. Eng. Technol.* **2013**, *2*, 3493–3499.
15. De La Ree, A.; Best, L.; Bradford, R.; Gonzalez-Arroyo, R.; Hepp, A. Fischer-Tropsch catalysts for aviation fuel production. In Proceedings of the 9th Annual International Energy Conversion Engineering Conference, San Diego, CA, USA, 31 July–3 August 2011; p. 5740.
16. Cano, F.M.; Gijzeman, O.; De Groot, F.; Weckhuysen, B. Manganese promotion in cobalt-based Fischer-Tropsch catalysis. *Stud. Surf. Sci. Catal.* **2004**, *147*, 271–276. [CrossRef]
17. Fan, G.; Li, F.; Evans, D.G.; Duan, X. Catalytic applications of layered double hydroxides: Recent advances and perspectives. *Chem. Soc. Rev.* **2014**, *43*, 7040–7066. [CrossRef]
18. Yu, J.; Wang, Q.; O'Hare, D.; Sun, L. Preparation of two dimensional layered double hydroxide nanosheets and their applications. *Chem. Soc. Rev.* **2017**, *46*, 5950–5974. [CrossRef]
19. Liao, P.; Zhang, C.; Zhang, L.; Yang, Y.; Zhong, L.; Wang, H.; Sun, Y. Higher alcohol synthesis via syngas over CoMn catalysts derived from hydrotalcite-like precursors. *Catal. Today* **2018**, *311*, 56–64. [CrossRef]
20. He, L.; Berntsen, H.; Ochoa-Fernández, E.; Walmsley, J.C.; Blekkan, E.A.; Chen, D. Co–Ni catalysts derived from hydrotalcite-like materials for hydrogen production by ethanol steam reforming. *Top. Catal.* **2009**, *52*, 206–217. [CrossRef]
21. Zhao, Q.; Ge, Y.; Fu, K.; Ji, N.; Song, C.; Liu, Q. Oxidation of acetone over Co-based catalysts derived from hierarchical layer hydrotalcite: Influence of Co/Al molar ratios and calcination temperatures. *Chemosphere* **2018**, *204*, 257–266. [CrossRef]

22. Hernández, W.Y.; Lauwaert, J.; Van Der Voort, P.; Verberckmoes, A. Recent advances on the utilization of layered double hydroxides (LDHs) and related heterogeneous catalysts in a lignocellulosic-feedstock biorefinery scheme. *Green Chem.* **2017**, *19*, 5269–5302. [CrossRef]
23. Kuśtrowski, P.; Chmielarz, L.; Bożek, E.; Sawalha, M.; Roessner, F. Acidity and basicity of hydrotalcite derived mixed Mg–Al oxides studied by test reaction of MBOH conversion and temperature programmed desorption of NH_3 and CO_2. *Mater. Res. Bull.* **2004**, *39*, 263–281. [CrossRef]
24. Xu, Y.; Wang, Z.; Tan, L.; Yan, H.; Zhao, Y.; Duan, H.; Song, Y.-F. Interface engineering of high-energy faceted Co_3O_4/ZnO heterostructured catalysts derived from layered double hydroxide nanosheets. *Ind. Eng. Chem. Res.* **2018**, *57*, 5259–5267. [CrossRef]
25. Aider, N.; Touahra, F.; Bali, F.; Djebarri, B.; Lerari, D.; Bachari, K.; Halliche, D. Improvement of catalytic stability and carbon resistance in the process of CO_2 reforming of methane by CoAl and CoFe hydrotalcite-derived catalysts. *Int. J. Hydrog. Energy* **2018**, *43*, 8256–8266. [CrossRef]
26. Maggi, R.; Martens, J.A.; Poncelet, G.; Grange, P.; Jacobs, P.A.; Delmon, B. *Preparation of Catalysts VII*; Elsevier Science: Amsterdam, the Netherlands, 1998.
27. Abelló, S.; Bolshak, E.; Montané, D. Ni–Fe catalysts derived from hydrotalcite-like precursors for hydrogen production by ethanol steam reforming. *Appl. Catal. A Gen.* **2013**, *450*, 261–274. [CrossRef]
28. Du, Y.-L.; Wu, X.; Cheng, Q.; Huang, Y.-L.; Huang, W. Development of Ni-based catalysts derived from hydrotalcite-like compounds precursors for synthesis gas production via methane or ethanol reforming. *Catalysts* **2017**, *7*, 70. [CrossRef]
29. Xiao, S.; Zhang, Y.; Gao, P.; Zhong, L.; Li, X.; Zhang, Z.; Wang, H.; Wei, W.; Sun, Y. Highly efficient Cu-based catalysts via hydrotalcite-like precursors for CO_2 hydrogenation to methanol. *Catal. Today* **2017**, *281*, 327–336. [CrossRef]
30. Gao, P.; Xie, R.; Wang, H.; Zhong, L.; Xia, L.; Zhang, Z.; Wei, W.; Sun, Y. Cu/Zn/Al/Zr catalysts via phase-pure hydrotalcite-like compounds for methanol synthesis from carbon dioxide. *J. CO2 Util.* **2015**, *11*, 41–48. [CrossRef]
31. Tanios, C.; Bsaibes, S.; Gennequin, C.; Labaki, M.; Cazier, F.; Billet, S.; Tidahy, H.L.; Nsouli, B.; Aboukaïs, A.; Abi-Aad, E. Syngas production by the CO_2 reforming of CH_4 over Ni–Co–Mg–Al catalysts obtained from hydrotalcite precursors. *Int. J. Hydrog. Energy* **2017**, *42*, 12818–12828. [CrossRef]
32. Izquierdo-Colorado, A.; Dębek, R.; Da Costa, P.; Gálvez, M.E. Excess-methane dry and oxidative reforming on Ni-containing hydrotalcite-derived catalysts for biogas upgrading into synthesis gas. *Int. J. Hydrog. Energy* **2018**, *43*, 11981–11989. [CrossRef]
33. Wierzbicki, D.; Baran, R.; Dębek, R.; Motak, M.; Grzybek, T.; Gálvez, M.E.; Da Costa, P. The influence of nickel content on the performance of hydrotalcite-derived catalysts in CO_2 methanation reaction. *Int. J. Hydrog. Energy* **2017**, *42*, 23548–23555. [CrossRef]
34. Wang, L.; Cao, A.; Liu, G.; Zhang, L.; Liu, Y. Bimetallic CuCo nanoparticles derived from hydrotalcite supported on carbon fibers for higher alcohols synthesis from syngas. *Appl. Surf. Sci.* **2016**, *360*, 77–85. [CrossRef]
35. Liao, P.-Y.; Zhang, C.; Zhang, L.-J.; Yang, Y.-Z.; Zhong, L.-S.; Guo, X.-Y.; Wang, H.; Sun, Y.-H. Influences of Cu Content on the Cu/Co/Mn/Al Catalysts Derived from Hydrotalcite-Like Precursors for Higher Alcohols Synthesis via Syngas. *Acta Phys.-Chim. Sin.* **2017**, *33*, 1672–1680. [CrossRef]
36. Forgionny, A.; Fierro, J.; Mondragón, F.; Moreno, A. Effect of Mg/Al Ratio on catalytic behavior of Fischer–Tropsch cobalt-based catalysts obtained from hydrotalcites precursors. *Top. Catal.* **2016**, *59*, 230–240. [CrossRef]
37. Jung, Y.-S.; Yoon, W.-L.; Seo, Y.-S.; Rhee, Y.-W. The effect of precipitants on Ni-Al_2O_3 catalysts prepared by a co-precipitation method for internal reforming in molten carbonate fuel cells. *Catal. Commun.* **2012**, *26*, 103–111. [CrossRef] [PubMed]
38. Sadek, R.; Chalupka, K.A.; Mierczynski, P.; Rynkowski, J.; Gurgul, J.; Dzwigaj, S. Cobalt based catalysts supported on two kinds of beta zeolite for application in Fischer-Tropsch synthesis. *Catalysts* **2019**, *9*, 497. [CrossRef]
39. Kang, S.-H.; Ryu, J.-H.; Kim, J.-H.; Jang, I.H.; Kim, A.R.; Han, G.Y.; Bae, J.W.; Ha, K.-S. Role of ZSM5 distribution on Co/SiO_2 Fischer–Tropsch catalyst for the production of C_5–C_{22} hydrocarbons. *Energy Fuels* **2012**, *26*, 6061–6069. [CrossRef]

40. Kurian, M.; Thankachan, S.; Nair, D.S.; Aswathy, E.K.; Babu, A.; Thomas, A.; Krishna, K.T.B. Structural magnetic, and acidic properties of cobalt ferrite nanoparticles synthesised by wet chemical methods. *J. Adv. Ceram.* **2015**, *4*, 199–205. [CrossRef]
41. Qiao, J.; Wang, N.; Wang, Z.; Sun, W.; Sun, K. Porous bimetallic $Mn_2Co_1O_x$ catalysts prepared by a one-step combustion method for the low temperature selective catalytic reduction of NO_x with NH_3. *Catal. Commun.* **2015**, *72*, 111–115. [CrossRef]
42. Liu, J.; Xiong, Z.B.; Zhou, F.; Lu, W.; Jin, J.; Ding, S.F. Promotional effect of H_2O_2 modification on the cerium-tungsten-titanium mixed oxide catalyst for selective catalytic reduction of NO with NH_3. *J. Phys. Chem. Solids* **2018**, *121*, 360–366. [CrossRef]
43. Kim, S.; Jeon, S.G.; Lee, K.B. High-temperature CO_2 sorption on hydrotalcite having a high Mg/Al molar ratio. *ACS Appl. Mater. Interfaces* **2016**, *8*, 5763–5767. [CrossRef]
44. Sun, L.; Yang, Y.; Ni, H.; Liu, D.; Sun, Z.; Li, P.; Yu, J. Enhancement of CO_2 adsorption performance on hydrotalcites impregnated with alkali metal nitrate salts and carbonate salts. *Ind. Eng. Chem. Res.* **2020**, *59*, 6043–6052. [CrossRef]
45. Wang, H.; Liu, W.; Wang, Y.; Tao, N.; Cai, H.; Liu, J.; Lv, J. Mg–Al Mixed oxide derived from hydrotalcites prepared using the solvent-free method: A stable acid–base bifunctional catalyst for continuous-flow transesterification of dimethyl carbonate and ethanol. *Ind. Eng. Chem. Res.* **2020**, *59*, 5591–5600. [CrossRef]
46. Wiyantoko, B.; Kurniawati, P.; Purbaningtias, T.E.; Fatimah, I. Synthesis and characterization of hydrotalcite at different Mg/Al molar ratios. *Procedia Chem.* **2015**, *17*, 21–26. [CrossRef]
47. Castaño-Robayo, M.-H.; Molina-Gallego, R.; Moreno-Guáqueta, S. Ethyl acetate oxidation over MnOx-CoOx. relationship between oxygen and catalytic activity. *CT & F-Cienc. Tecnol. Futuro* **2015**, *6*, 45–56.
48. Gong, K.; Lin, T.; An, Y.; Wang, X.; Yu, F.; Wu, B.; Li, X.; Li, S.; Lu, Y.; Zhong, L.; et al. Fischer-Tropsch to olefins over CoMn-based catalysts: Effect of preparation methods. *Appl. Catal. A Gen.* **2020**, *592*, 117414. [CrossRef]
49. Cui, M.; Hou, Y.; Zhai, Z.; Zhong, Q.; Zhang, Y.; Huang, X. Effects of hydrogen peroxide co-precipitation and inert N_2 atmosphere calcination on CeZrLaNd mixed oxides and the catalytic performance used on Pd supported three-way catalysts. *RSC Adv.* **2019**, *9*, 8081–8090. [CrossRef]
50. Cai, L.-N.; Guo, Y.; Lu, A.-H.; Branton, P.; Li, W.-C. The choice of precipitant and precursor in the co-precipitation synthesis of copper manganese oxide for maximizing carbon monoxide oxidation. *J. Mol. Catal. A Chem.* **2012**, *360*, 35–41. [CrossRef]

© 2020 by the authors. Licensee MDPI, Basel, Switzerland. This article is an open access article distributed under the terms and conditions of the Creative Commons Attribution (CC BY) license (http://creativecommons.org/licenses/by/4.0/).

Article

Tuning the Selectivity of LaNiO$_3$ Perovskites for CO$_2$ Hydrogenation through Potassium Substitution

Constantine Tsounis, Yuan Wang †, Hamidreza Arandiyan *,‡, Roong Jien Wong §, Cui Ying Toe, Rose Amal and Jason Scott *

Particles and Catalysis Research Group, School of Chemical Engineering, The University of New South Wales, Sydney, NSW 2052, Australia; c.tsounis@unsw.edu.au (C.T.); yuan.wang4@unsw.edu.au (Y.W.); roong.jien.wong@rmit.edu.au (R.J.W.); c.toe@unsw.edu.au (C.Y.T.); r.amal@unsw.edu.au (R.A.)

* Correspondence: hamid.arandiyan@sydney.edu.au (H.A.); jason.scott@unsw.edu.au (J.S.); Tel.: +61-2-9114-2199 (H.A.); +61-2-9385-7361 (J.S.)
† Permanent Address: School of Chemistry, The University of New South Wales, Sydney, NSW 2052, Australia.
‡ Permanent Address: Laboratory of Advanced Catalysis for Sustainability, School of Chemistry, The University of Sydney, Sydney, NSW 2006, Australia.
§ Permanent Address: Applied Chemistry and Environmental Science, School of Science, RMIT University, Melbourne, Victoria 3000, Australia.

Received: 26 February 2020; Accepted: 7 April 2020; Published: 8 April 2020

Abstract: Herein, we demonstrate a method used to tune the selectivity of LaNiO$_3$ (LNO) perovskite catalysts through the substitution of La with K cations. LNO perovskites were synthesised using a simple sol-gel method, which exhibited 100% selectivity towards the methanation of CO$_2$ at all temperatures investigated. La cations were partially replaced by K cations to varying degrees via control of precursor metal concentration during synthesis. It was demonstrated that the reaction selectivity between CO$_2$ methanation and the reverse water gas shift (rWGS) could be tuned depending on the initial amount of K substituted. Tuning the selectivity (i.e., ratio of CH$_4$ and CO products) between these reactions has been shown to be beneficial for downstream hydrocarbon reforming, while valorizing waste CO$_2$. Spectroscopic and temperature-controlled desorption characterizations show that K incorporation on the catalyst surface decrease the stability of C-based intermediates, promoting the desorption of CO formed via the rWGS prior to methanation.

Keywords: selectivity tuning; CO$_2$ methanation; reverse water gas shift

1. Introduction

An economical yet effective removal and utilization of CO$_2$ has not been implemented on a wide scale level to reduce its impact on climate change through the greenhouse effect. This makes the efficient methanation of CO$_2$ or the production of CO through the reverse water gas shift (rWGS) reaction a promising solution if it can be implemented viably [1,2]. Current major solutions of CO$_2$ utilization revolve around the production of synthetic chemicals such as methanol, various carbonates, and urea. Combined with other carbon sequestration efforts, this only accounts for the removal of a minute portion of the ~ 30 gigatonnes (Gt) of CO$_2$ emitted annually [3–5]. The production of these liquid chemicals generally require intensive operating conditions, and reactions are considerably slower than the rWGS or Sabatier methanation reactions [1]. Furthermore, there is a significant global push towards realizing renewable hydrogen production at scale through the use of water electrolysis combined with renewable energy inputs [2]. This presents significant opportunity for the implementation of CO$_2$ hydrogenation through these routes, which are able to convert CO$_2$ into CO (via the rWGS) and CH$_4$ (via methanation).

CH$_4$ and CO produced from the thermal reduction of CO$_2$ can be used to create sustainably sourced synthetic fuels, as well as act as building blocks for other reactions which are able to produce value added products. These include aromatic compounds, alcohols, ketones, and carboxylic acids. Moreover, the process of creating these products can reduce anthropogenic CO$_2$ in the atmosphere, presenting itself as a potential multifaceted approach to reducing CO$_2$ pollution [4].

CO$_2$ + 4H$_2$ → CH$_4$ + 2H$_2$O	Sabatier Reaction [6]	$\Delta H = -165.0$ kJ mol^{-1}
CO$_2$ + H$_2$ → CO + H$_2$O	Reverse Water Gas Shift [6]	$\Delta H = 41.15$ kJ mol^{-1}

Further to this, there is a need for specific product ratios of CH$_4$ and CO, which are able to be used as reactant streams for further hydrocarbon reforming. It has been shown previously that the addition of CO during the dehydrocondensation of CH$_4$ into aromatic products over a zeolite-based catalyst was able to significantly improve catalytic stability and selectivity, resulting in an increase in catalytic performance [7–9]. Therefore, in order to increase the feasibility and applicability of utilizing the catalytic conversion of CO$_2$ into sustainable fuels and products, this work aimed to tune the selectivity of CO$_2$ reduction between the rWGS and methanation to specific product ratios, which can be used more efficiently for downstream processes. Perovskite-type oxides with the general formula ABO$_3$ containing both rare earth elements and 3d transition metals have received much attention in recent years [10–13]. Supported base metal oxides (e.g., cerium oxide [14] and zirconium oxide [15]) and perovskites such as LaMO$_3$ (M = Co, Ni, Mn) [16] are catalytically active in CO$_2$ reduction at above 200 °C, while showing good selectivity towards the Sabatier reaction. Particularly, Ni as the B-site cation has been previously probed for CO$_2$ methanation. In using LaNiO$_3$ (LNO), which is reduced to Ni/La$_2$O$_3$ upon activation [17,18], the Ni species have been shown to be an active component in both Sabatier methanation and the rWGS routes, and, under various conditions, will have different reaction selectivity and activity [19,20]. However, very few works have been reported to exploit the strong thermal, hydrothermal, excellent redox properties, and flexible composition of the perovskite structure [12,21,22], and specifically manipulate selectivity away from CO$_2$ hydrogenation to the rWGS pathways, achieving an optimal ratio of CO and CH$_4$ in the product stream.

Alkali metals such as Li, Na, and K within various metal-based supported catalysts have been shown to promote rWGS routes due to their role of forming active sites, which promote formate species that decompose into CO. These active sites also alter the properties of neighboring species by potentially changing stability, particle size, and intermediate bond types [23–26]. Furthermore, K has been shown previously to be readily incorporated into La-based perovskites through the A-site substitution of La, and is able to change its electronic properties due to differences in ionic radius and preferred oxidation states [27,28].

In this work, we exploited the high selectivity of the LNO perovskite catalyst towards the Sabatier reaction and, by controlling potassium incorporation in the catalyst to form La$_{1-x}$K$_x$NiO$_3$ (x= 0, 0.1, 0.2, or 0.3, denoted as 0K LNO, 10K LNO, 20K LNO, or 30K LNO), tuned the reaction pathway toward the rWGS routes. The resulting changes in morphology, electronic properties, reducibility, and surface compositions emerging from the K substitution were analyzed in terms of their effect on the selectivity of the catalysts.

2. Results and Discussion

2.1. Material Characterization

The XRD pattern (Figure 1) for the LNO perovskite without K substitution showed high crystallinity and diffraction peaks that corresponded to the rhombohedral LaNiO$_3$ perovskite phase (JCPDS 00-034-1181). It was observed that after partial substitution of La with K, a secondary major phase appeared corresponding to the formation of NiO (JCPDS 01-078-0643) in conjunction with the crystalline LaNiO$_3$ structure. A similar phenomenon was obtained in other works with the La$_x$K$_{1-x}$CoO$_3$ (x = 0–0.3) perovskite, whereby K substitution induced the formation of the major B-cation oxide (Co$_3$O$_4$) [27].

It was postulated that, in our case, a structural defect occurred as a result of the different ionic radii between La^{3+} and K^+, resulting in an imperfect substitution, where not all La^{3+} ions were replaced by K^+ within the structural A-site. This may result in an excess of Ni species not incorporated into the perovskite structure, which react with oxygen to form NiO during calcination. The NiO species formed increased in abundance as K substitution increased, as seen in Figure 1 through the increasing XRD peak intensity of NiO. Slight peak broadening in the characteristic peak of the LNO perovskite also indicated lattice distortion, potentially due to the substitution of K cations into the structure. However, the presence of K species could not be detected in XRD analysis, suggesting that the K in the catalyst was in a well-dispersed form, such as nanocrystals [26].

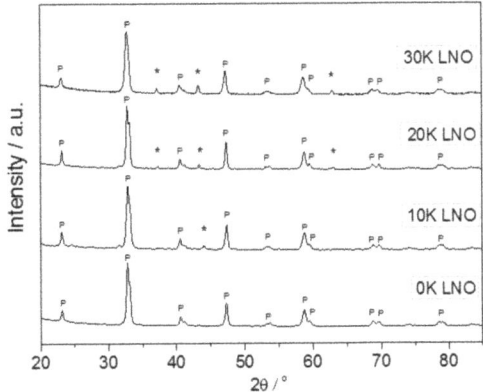

Figure 1. Characteristic peak intensity normalized XRD patterns of 0–30K LNO catalysts. The (*) symbol represents the NiO standard phase while (P) symbol represents the $LaNiO_3$ perovskite phase.

The Brunauer-Emmett-Teller (BET) surface area for all catalysts are shown in Table 1. The 0K LNO possessed the highest surface area value of 10.5 $m^2\ g^{-1}$. Previous works reported surface areas of $LaNiO_3$ with various synthesis methods ranging from approximately 1–10 $m^2\ g^{-1}$ [29,30]. For the 10–30K LNO samples, the BET surface area was lower following La substitution with the K ions. This was partially attributed to the distortion of the rhombohedral lattice structure of the single phase $LaNiO_3$ due to a change in radius of the A-site cation, and has also been correlated to the formation of low melting point K compounds, as shown by Xu and coworkers [27], which were deformed during the calcination.

Table 1. Surface parameters of the 0–30K LNO samples.

Catalyst	BET Surface Area ($m^2\ g^{-1}$)	K Surface Amount [a,b] (at. %)	Overall K Amount [b,c] (at. %)
0K LNO	11	0	0
10K LNO	6.5	30.4	0.56
20K LNO	5.2	23.8	1.3
30K LNO	7.1	3.56	2.4

[a] Determined using XPS surface analysis. [b] Calculated based on percentage of overall metal content. [c] Calculated through ICP-OES analysis.

N_2 adsorption and desorption curves showed a small hysteresis indicating adsorbed N_2 on the nanoparticle surface for the 0K LNO sample, which indicated the limited presence of meso or micropores [31], as can be seen in Figure S1. These pores are beneficial to the overall catalytic performance as they allow for higher adsorbed surface oxygen capabilities and better redox properties. The hysteresis appeared to decrease as K was incorporated into the catalyst structure, providing further evidence of distortion within the crystal lattice due to the K substitution amount. Interestingly, Table 1 shows that as higher amounts of K were incorporated during synthesis there was a decrease in K surface species, suggesting that the substitution effect was more evident for higher K amounts in our system.

For reductive type reactions, catalyst reducibility is a good indicator of the redox properties and abilities over a temperature range in which the catalyst is active. Particularly for metal oxide-based catalysts, the formation and abundance of active sites is a driving force for the catalytic reduction of CO_2, as well as a factor defining reaction pathway [17]. Figure 2 presents the reductive potential through H_2 temperature programmed reduction (H_2 TPR) of the 0–30 K LNO catalysts. The results show two major reduction zones across a temperature range of approximately 250–575 °C. Singh et al. described two possible reaction paths for $LaNiO_3$ reduction, which were distinguishable through the relative sizes of the first (α) and second (β) reduction peak. However, both pathways eventually led to solid phase crystallization, forming Ni supported on La_2O_3 [17]. For the crystalline 0K LNO, whereby the only phase present exists as a rhombohedral $LaNiO_3$ perovskite, it can be seen that the β peak is approximately two times the size of the α peak. The pathway likely followed a two-step method, whereby the α peak indicates the reduction of Ni^{3+} to Ni^{2+}, and the β peak indicates the exsolution of Ni onto the perovskite surface through the reduction of Ni^{2+} to Ni^0, forming an active Ni site, which has been shown to participate in CO_2 methanation routes [19,32].

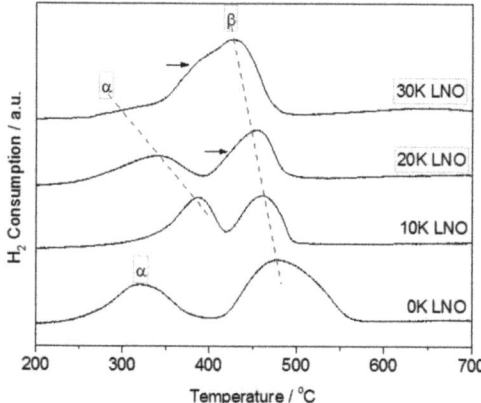

Figure 2. H_2 TPR profiles of the 0–30K LNO samples.

In the case of 10–30K LNO, a similar reduction pathway was proposed. However, slight changes in the peak profiles were seen, likely as a result of an increase in structural defects as K was introduced. It can be seen that the α peak initially shifted toward higher temperatures, the extent of which may be directly influenced by the amount of K surface-based species. The effect tended to reduce for the 20–30K LNO samples as the peaks shifted back towards lower temperatures. This suggests that surface K species may increase the stability of Ni^{3+}, as seen by their increased resistance to reducibility [17]. Conversely, the Ni^{2+} to Ni^0 reduction peak shifted slightly to lower temperatures as K concentrations in the bulk catalyst increased, which may result from perovskite lattice distortion due to changes in the A-site ionic radius. Additionally, a tailing peak, which developed on the left side of the β peak in the 20K LNO sample, increasing into a shoulder on the 30K LNO sample, may be ascribed to the reduction of NiO, which may not have entered the perovskite after substitution. The increasing NiO crystal size indicated by the proportionally larger NiO XRD diffraction peaks (Figure 1) may create a slightly more easily reduced Ni^{2+} phase, which can be seen by the increasing shoulder peak at lower temperatures in these samples. The decreasing area of the α and β peaks in the 10–30K LNO samples correlated to the presence of less reducible Ni species once K was incorporated into the structure and surface. This may be ascribed to K hindering the abundance of active Ni species that participate in redox reactions, and possibly influenced hydrogen activation on the catalyst [20].

TEM images of the 20K LNO sample shown in Figure 3a–c reveal NiO particles were formed on the perovskite surface during synthesis. There appears to be an area of high contrast surrounding the Ni species, which may possibly be ascribed to K species on the catalyst surface, which are likely in oxide form [33]. EDS mapping of the 20K LNO sample, as shown in Figure 3d–h, further confirms the presence of surface K species, in conjunction with La, Ni, and O species also present on the catalyst surface. When comparing 0K LNO and 20K LNO after the reduction treatment, prior to performance evaluation, TEM images demonstrated a significant increase in particle size and decrease in uniform distribution of reduced Ni nanoparticles in the 20K LNO sample relative to the 0K LNO sample (Figure S2). This effect was likely due to the reduction of bulky NiO species previously formed on the 20K LNO surface during synthesis, as La was replaced by K cations.

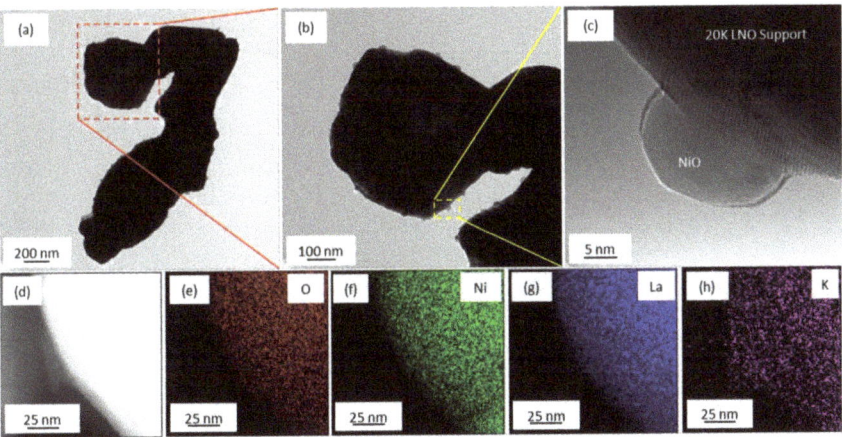

Figure 3. (**a–c**) TEM images of the fresh 20K LNO catalyst and (**d–h**) EDS mapping of O, Ni, La, and K on the surface of the sample.

XPS analysis was used to analyze the changing chemical properties of the samples as K ions were substituted into the perovskite structure. Figure 4a depicts two main deconvoluted peaks of K, the first of which at approximately 292.6 eV corresponded to the $2p_{3/2}$ orbital, with an accompanying $2p_{1/2}$ orbital peak at approximately 295.4 eV. These peaks were shown to correspond to a K–O group, which, in our case, likely consisted of K_2O which formed through the calcination of K surface species in air [33–35]. Observation showed a shift to higher binding energies for both of these peaks to 293.0 eV and 295.8 eV in the 30K LNO sample for the $2p_{3/2}$ and $2p_{1/2}$ orbitals, respectively, indicating oxidation of the surface K cation sites as more K was substituted into the perovskite structure. Furthermore, the decreasing intensity of these peaks in the 30K LNO sample was a consequence of the decreasing amount of surface K, shown in Table 1.

For the Ni 2p orbitals shown in Figure 4b, the 0K LNO sample exhibited two main peaks at approximately 855.9 eV and 873.0 eV, which were previously reported to correspond to the $2p_{3/2}$ and $2p_{1/2}$ orbitals of Ni^{3+} in the perovskite structure, respectively [36]. It can be seen that these peaks shifted towards lower binding energies in the 10K and 20K LNO samples, indicating that Ni species were being reduced, the electron source of which likely originated from both surface and bulk K species. Further evidence of the reduction effect was seen through a shoulder peak at approximately 853.9 eV in the 0K LNO sample, which increased in intensity for the 10K and 20K LNO samples, indicative of the formation of Ni^{2+} species. The shoulder, however, became much more pronounced, forming a full peak in the 30K LNO sample, confirming the presence of the Ni^{2+} species facilitated by electron donation from K–O species. Upon reduction during the reaction, it was seen that all Ni^{3+} species were fully reduced to Ni^0, which as previously mentioned, formed an active species for CO_2 hydrogenation (Figure S3 (reduced sample Ni 2p XPS spectra)).

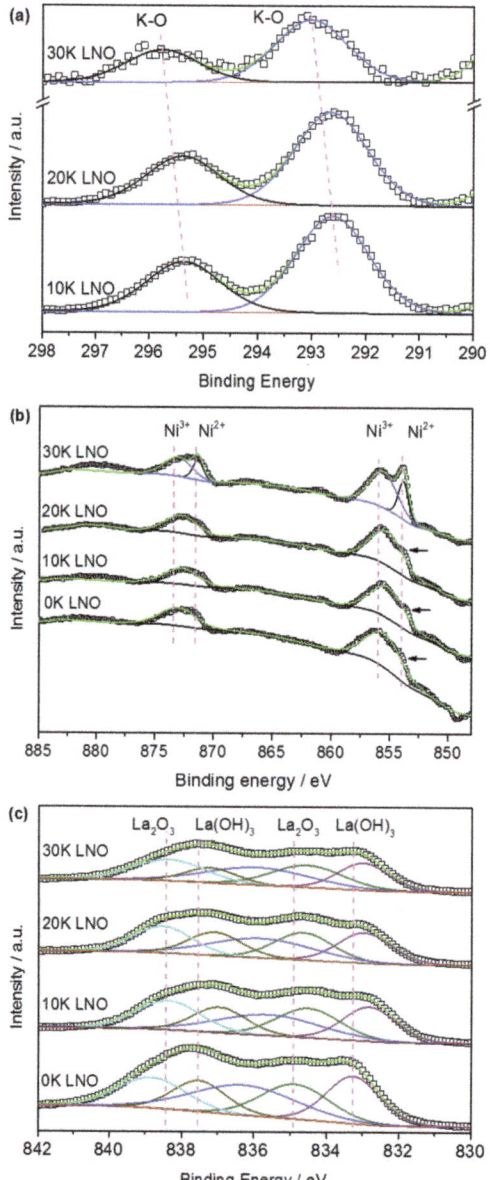

Figure 4. XPS profiles of (a) K 2p, (b) Ni 2p, and (c) La 3d of 0–30 K LNO samples.

The La 3d region was also examined and deconvoluted into two main doublets corresponding to La_2O_3 and $La(OH)_3$, whose spectra exhibited both core and satellite peaks. Peak overlap between $La(OH)_3$ and $La_2(CO_3)_3$ indicated that there was also $La_2(CO_3)_3$ present when considering carbonate species present in the C1s spectra (Figure S4). Figure 4c shows the deconvolution of the La $3d_{5/2}$ orbital where the La_2O_3 core level and satellite peak were located at 833.2 eV and 837.5 eV (ΔE = 4.3 eV), while $La(OH)_3$ core level and satellite peaks were found to be at 834.9 eV and 838.9 eV (ΔE = 4.0 eV), respectively [37]. The peaks were observed to shift towards lower binding energies; however, as the amount of K substitution increased, the peaks drifted back towards slightly higher energies. It was proposed that the electron source of the La_2O_3 and $La(OH)_3$ reduction originated primarily from surface rather than bulk K species, and, therefore, the effect slightly decreased as the surface K concentration reduced to 3.56 at. % in the 30K LNO sample. Substituted K, in contrast to its effects on Ni, is believed to have minimal or negligible effect on the La species binding energies (electron densities) as K within the perovskite structure were not within the first shell vicinity of La, i.e., not having direct interaction with La. Despite the slight drift back to higher binding energies, the core level binding energies for La_2O_3 and $La(OH)_3$ in the 30K LNO sample still decreased overall by 0.4 eV and 0.3 eV, respectively, compared to 0K LNO, and may have effected intermediate bond stability during the reaction.

2.2. Catalytic Performance

2.2.1. Carbon Dioxide Conversion and Reaction Selectivity

The catalytic activity was observed at atmospheric pressure, over a temperature range that yielded from 0% CO_2 conversion to 100%, based on catalysts which were reduced in a flow of H_2 (25 mL min^{-1}) at 500 °C for 2 h. The reduction method was used to promote the formation of Ni^0, as Ni^0 is known to be an active species in both methanation and rWGS reactions [19,20]. The results from the catalytic activity test are shown in Figure 5 where CO_2 conversion is displayed as a function of temperature (light off curves). The highest performing catalyst, 0K LNO, exhibited CO_2 conversion with $T_{50\%}$ (temperature at which 50% of CO_2 is converted) of 240 °C, and $T_{100\%}$ at 270 °C, a performance comparable to some of the highest performing catalysts for the methanation reaction under similar reaction conditions [1]. For the K-substituted LNO catalysts, a trend in decreasing CO_2 conversion at a given temperature was identified and can be attributed to the incorporation of K. For the 10K, 20K, and 30K LNO catalysts, the $T_{50\%}$ temperatures are 290 °C, 306 °C, and 316 °C, respectively. The decrease in catalytic activity can be explained as a result of lesser stable adsorbed CO_2 on surface K, (shown to be in an intermediate K_2CO_3 state under in situ conditions), and the hindrance of activated H spill over from Ni species to the adsorbed CO_2 that K species may cause [33]. Furthermore, when comparing 10K LNO and 30K LNO in Figure 4c, where fewer electrons from K are transferred to La sites, the relatively lower electron density may decrease CO_2 adsorption and H activation rates on the catalyst. This effect, combined with a slight decrease in surface area shown in Table 1, may further lower overall catalytic activity. Moreover, the poor Ni distribution in the K-substituted samples and increased particle size (Figure S2) may also contribute to their decrease in CO_2 conversion rates [38]. Despite this, stability tests showed that for both 0K LNO and 20K LNO catalysts, over a period of 5 h, reaction activity and selectivity remained constant (Figure S5).

Reaction selectivity for all catalysts was observed to remain between the methanation of CO_2 (CH_4 production) and the reverse water gas shift (CO production), as seen in Figure 6. The 0K LNO can be seen as a high performing catalyst for the Sabatier methanation reaction, with 100% selectivity towards the methanation reaction for all temperatures. The increasing incorporation of K species within the perovskite structure can be used to tune the selectivity from 100% methanation at a given temperature, to also include CO in the product stream depending on catalyst and operating temperature, as seen in Figure 6. The reaction tuning can also be seen in Figure 7, which demonstrated the selectivity of all catalysts when overall CO_2 conversion was at 50%, showing that the changes in selectivity as a result of the addition of K were independent of overall conversion between the catalysts.

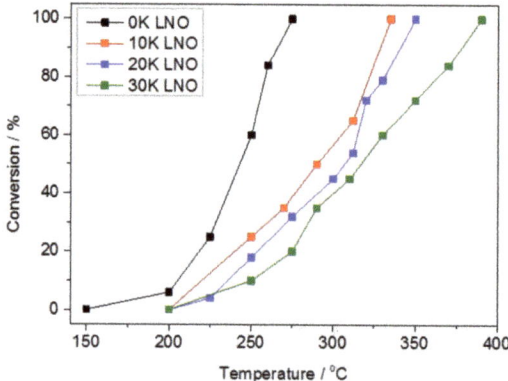

Figure 5. CO_2 conversion as a function of temperature for the 0–30K LNO samples.

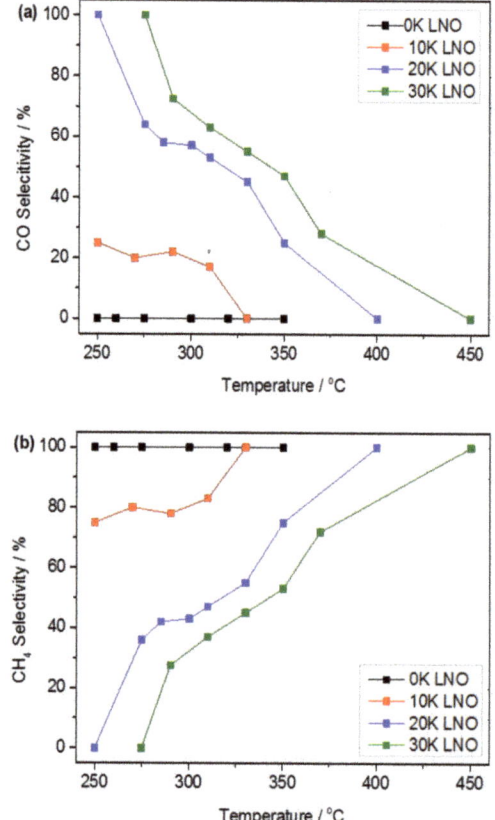

Figure 6. (**a**) Reaction selectivity towards the rWGS reaction and (**b**) reaction selectivity towards Sabatier methanation for all samples for 0–100% conversion.

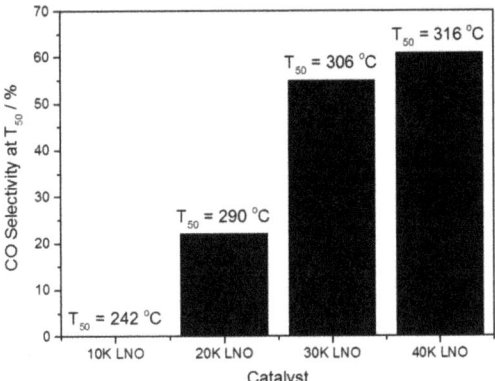

Figure 7. Selectivity towards the rWGS at $T_{50\%}$.

The selectivity of transition metal catalysts is largely influenced by the metal-support interaction, as well as the adsorption and stability of intermediate species on the support, which determine the most thermodynamically favorable reaction pathway [39]. For example, it has been reported that Ni/Ce-Zr-O-supported catalysts can exhibit up to 100% selectivity towards the rWGS reaction pathway [40], while Ni/CeO$_2$ can attain 100% selectivity towards CO$_2$ methanation [41]. Other transition metal catalysts also show a similar phenomenon with varying reaction selectivity [42–44]. Therefore, it is likely that, by incorporating K species, support interactions with the Ni active sites as well as the intermediate species stability on the support can be systematically tuned. This apparent shift in selectivity in K-substituted samples relative to 0K LNO may be partially contributed to by the increase in size of Ni particles (Figure S2). However, to fully understand the reasons for selectivity towards the rWGS, investigation into the role of K substitution was undertaken for the 0K LNO and 20K LNO samples.

2.2.2. Role of Potassium Substitution in Changing Reaction Selectivity

CO temperature programmed desorption (CO-TPD) was undertaken to understand the nature of CO adsorption/desorption equilibrium on the catalyst surface. Figure 8a outlines the CO desorption results of the 0K and 20K samples. It is clear that there were a number of CO desorption features that were identifiable for the two samples. The first peak for both samples (labelled as α at 264 °C in 0K LNO and 164 °C in 20K LNO) corresponded to CO species adsorbed on Ni^{2+} [45]. It is likely that surface adsorption also occurred on the K$_2$O surface species for 20K LNO, forming a range of intermediates such as carbonates and hydroxycarbonates, resulting in the shift of the first CO desorption peak to a lower temperature with lower intensity, indicating reduced CO uptake and stability. A second peak, present clearly only for 0K LNO at 535 °C (labelled as β), may be indicative of CO species adsorbed on Ni^0 particles [45]. Although the formation of Ni carbonyl complexes may also occur, these complexes are easily removed at low temperature <100 °C and do not form a significant part of the desorption profile [46]. The lack of obvious peak in this region for the 20K LNO sample was further evidence of the poorer stability and interaction of adsorbed CO on this sample. Two sharp desorption peaks in the 20K LNO sample were present in the region of 600 °C to 900 °C. These peaks may derive from the thermal decomposition of unstable K$_2$CO$_3$ species and have relatively high intensity due to the presence of other decomposition products, such as CO$_2$, KOH, and KHCO$_3$ [47].

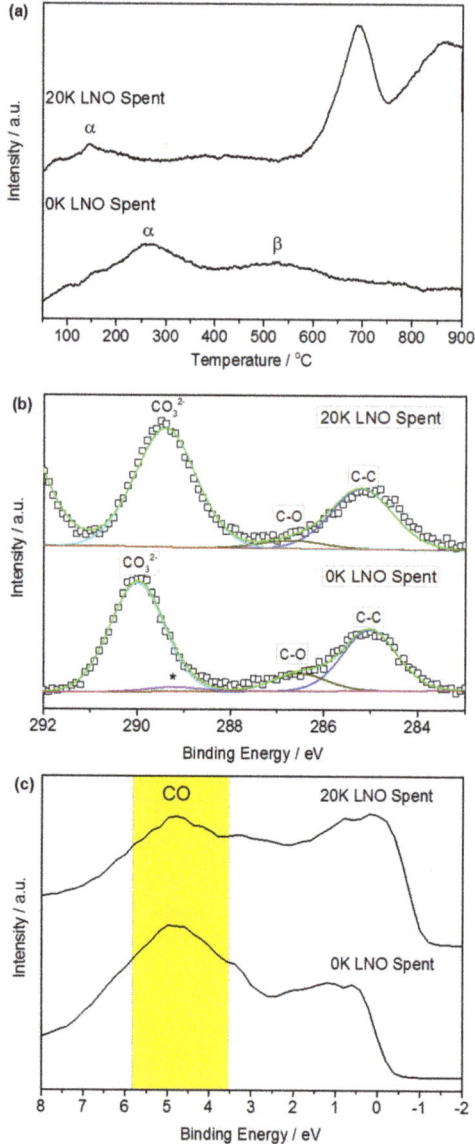

Figure 8. (a) CO-TPD profiles of the 0K and 20K LNO samples. (b) C1s spectra of the spent 0K and 20K LNO samples after a 2-h reduction step and 5-h reaction period at 300 °C. (c) XPS valence band spectra of the same spent samples.

In the presence of K_2O, it was shown by Maneerung et al. that CO_2 adsorption may take place on K_2O surface species, leading to the formation of K_2CO_3 [33]. This intermediate was shown to be inherently more unstable when compared to $La_2O_2CO_3$ intermediates, which form as CO_2 is adsorbed on the La surface species. Therefore, when taken with the CO-TPD, it was proposed that the K incorporation is more likely to promote the formation of less stable C-based intermediates, which may preferentially desorb when compared to C-based intermediates in the analogous 0K LNO sample. Traditionally, one of the CO_2 methanation mechanisms has been attributed to CO_2 adsorption and dissociation to form CO (via the rWGS), followed by CO hydrogenation [19,48,49]. It is possible that, in the presence of K species on the surface of our catalysts, these CO intermediates avoided complete hydrogenation before desorbing from the surface, forming CO instead of CH_4. Consequently, K incorporation and the associated surface species were able to tune reaction selectivity when comparing pristine LNO and K-substituted LNO samples, as seen in Figure 7.

Further evidence of C-based intermediate stability can be seen in Figure 8b, which provides the C1s XPS spectra of the spent 20K and 0K LNO samples, after a two-hour reduction in H_2 and five-hour reaction step. A peak with binding energy 289.2 eV (indicated with an asterisk) can be observed in the 0K LNO sample, corresponding to the formation of C=O species adsorbed on the catalyst surface [50]. In the case of the 20K LNO sample, the absence of this peak may indicate the evacuation of carbonate species during the reaction due to instability induced by the K cations forming carbonates or hydroxycarbonates, indicative of the shifting reaction selectivity of the K-substituted samples through promotion of rWGS routes. A similar phenomenon was observed for the C-O peak at 286.5 eV [50], which decreased in intensity with K incorporation (20K LNO), further alluding to the decrease in carbon-oxide species on the 20K LNO sample due to their increased instability during reaction.

In support of the role of K in shifting the reaction selectivity towards rWGS routes due to premature desorption of CO intermediates, analysis of the valence band spectra was undertaken in the spent samples, as shown in Figure 8c. It is apparent that at ca. 5 eV there was an increase in peak intensity for the 0K LNO when compared to the 20K LNO spent samples. This can be attributed to the presence of CO adsorbed on the catalyst surface due to overlapping of the CO valence band with the overall valence electronic structure of the catalyst surface, indicative of the heightened C-intermediate stabilities in the absence of K surface species [51]. This effect is demonstrated in Figure S6, which illustrates a proposed mechanism in which K_2O species promote the instability of C-based intermediates, based on previous studies [33].

XPS analysis of the spent catalysts (Figure S7) showed the binding energy of the K 2p peak in the 20K sample to be at 292.9 eV with a split-orbit peak at 295.6 eV, compared to 292.6 eV and 295.4 eV, respectively, in the fresh 20K LNO sample. This indicates that as Ni species were reduced, K species were further oxidized, suggesting that K played the role of an electron source for this reduction step. Consequently, the oxidation state of K may be an underlying reason behind the shifting selectivity due to changes in intermediate stability [52]. The proposed K to Ni electronic pathway agrees with the work function-driven electron transfer mechanism observed in our earlier works on bimetallic alloys [53,54], resulting in the decreased H_2 TPR peak for the reduction of Ni^{2+} into Ni^0.

3. Materials and Methods

3.1. Catalyst Preparation

A citrate method similar to that described by Irusta et al. [55] was used to create the nominal chemical compositions of $LaNiO_3$, $La_{0.9}K_{0.1}NiO_3$, $La_{0.8}K_{0.2}NiO_3$, and $La_{0.7}K_{0.3}NiO_3$ catalyst samples. The samples were denoted as 0K LNO, 10K LNO, 20K LNO, and 30K LNO, respectively. Each of the metal salts, $La(NO_3)_3 \cdot 6H_2O$ (>99%, Sigma-Aldrich®, St. Louis, MO, USA), $Ni(NO_3)_2 \cdot 6H_2O$ (>99%, Ajax Finechem, Scoresby, VIC, Australia), and KNO_3 (>99%, Sigma-Aldrich®) were dissolved in stoichiometric amounts in 4 mL of distilled water (Milli-Q, 18 mΩ cm). Citric acid (>99%, Sigma Aldrich ®) was added in a molar ratio such that for every one mole of the metal ion dissolved in solution two moles of citric acid were added. The dissolved metal ion solution was magnetically stirred for one hour at ambient temperature, forming a homogenous mixture. The solution was dried at 120 °C to give a porous gel, which was crushed to obtain a homogenous powder. The as-prepared solid was calcined in air for 3 h at 750 °C, being heated at a rate of 5 °C min^{-1}.

3.2. Catalyst Characterization

The crystallinity of each sample was analyzed by X-ray diffraction (XRD) using a Philips PANalytical Scherrer Diffractometer (λ= 0.154 nm, Amsterdam, Netherlands). Scattering intensity was recorded at a range of 8° < 2θ < 90° with a 2θ step of 0.03° and a time of 2 s per step. Hydrogen temperature programmed reduction (H_2 TPR) was performed using a Micromeritic Autochem II 2920 (Norcross, GA, USA) equipped with a thermal conductivity detector (TCD). Approximately 50 mg of the sample were first heated to 150 °C at a rate of 10 °C min^{-1} in 40 mL min^{-1} of argon, and held for 30 min to remove moisture. The temperature was then reduced to 50 °C, where the reducing gas was introduced at 40 mL min^{-1} (10% H_2 in Ar). The sample was then heated to 900 °C at 10 °C min^{-1}, while H_2 consumed passing through the catalyst bed was measured through the TCD. CO temperature programmed desorption (CO-TPD) was also undertaken on the same apparatus. 50 mg of the fresh samples were initially treated with 30 mL min^{-1} of He where it was heated to 150 °C, at a rate of 10 °C min^{-1}, and held for 30 min to remove moisture. The sample was then cooled to 50 °C where CO was introduced at a flow rate of 20 mL min^{-1} for 30 min, followed by He at a flow rate of 20 mL min^{-1} for 60 min. The temperature was then increased to 900 °C at a rate of 10 °C min^{-1} while any desorbed CO in the effluent gas was detected through the TCD. BET surface area, pore volume, and pore size of the samples were measured using N_2 physisorption on a Micrometric Tristar 3030. Samples were pretreated under vacuum at 150 °C for 3 h to remove impurities. X-ray photoelectron spectroscopy (XPS) analysis was undertaken using a Thermo Fisher model ESCALAB250Xi (Waltham, MA, USA) and was used to probe the chemical states of various surface species with a wave energy (Al Kα) of 1486.68 eV and C1s reference of 284.8 eV. For the spectra of the spent samples, the samples were kept away from ambient oxygen by flushing the reactor with N_2, prior to storage in sealed N_2-purged vials before measurement. Morphology of the fresh catalyst samples was characterized using field-emission high resolution transmission electron microscopy (FE-HRTEM) on a Philips CM200 microscope (Amsterdam, Netherlands). Surface composition was analyzed using a high-angle annular dark-field scanning transmission electron microscope energy-dispersive X-ray spectroscopy (HAADF-STEM-EDS) with a JEOL F200 STEM (Tokyo, Japan). Chemical composition of the catalysts was determined through inductively coupled plasma optical emission spectrometry (ICP-OES) analysis and was undertaken on fresh samples that were digested in 1 part HNO_3 (70% w/w) and 3 parts HCl (36% w/w).

3.3. Catalytic Apparatus and Reaction

The performance of the catalysts was evaluated on a rig consisting of a quartz tube (6 mm inner diameter) acting as a fixed-bed micro reactor (see Figure S8 for a detailed schematic). Then, 50 mg of the sample was loaded on a bed of quartz wool. The samples were first reduced for 2 h at 500 °C in a 25 mL min^{-1} stream of H_2 gas. The temperature was then lowered to 200 °C where N_2 at 13 mL min^{-1} and CO_2 at 2 mL min^{-1} were introduced and stabilized for 45 min, giving a total gas hourly space velocity (GHSV) of 48,000 mL (g h)$^{-1}$. The reactant gas was passed through the catalyst, with the exiting gas passing through a gas chromatograph (Young Lin-6100, Gyeonggi-do, Korea) containing a Carboxen-1010 PLOT column, equipped with a thermal conductivity detector (TCD). The outlet gas composition was analyzed over a temperature range which allowed for 0–100% CO_2 conversion of all catalysts.

4. Conclusions

The findings demonstrated the capacity for facile tuning of the selectivity in high-performing $LaNiO_3$ methanation catalysts towards the rWGS reaction. Tunable selectivity can potentially be beneficial for downstream hydrocarbon reforming while valorizing CO_2, providing a clear motivation for this method of rational catalyst design. Incorporating K into the catalytic structure produced K surface species, which were proposed to be the main contributor to the effect, by promoting the formation of thermally unstable C-intermediates, which potentially escape hydrogenation and instead dissociate via the rWGS. The effect was demonstrated by comparing CO interaction with the catalysts, whereby 20K LNO samples showed less CO adsorption uptake and increased instability. Analysis of spent catalysts alluded to the presence of C-O and C=O based species, which had desorbed from the K-substituted samples while remaining for the 0K LNO samples, further evidence of the increased instability of C-intermediates in K-substituted samples. The results were confirmed by analyzing the valence band spectra of the samples, which showed a clear increase in binding energy intensity at 5 eV for the 0K LNO sample, corresponding to the presence of CO residue.

Supplementary Materials: The following are available online at http://www.mdpi.com/2073-4344/10/4/409/s1. Figure S1: N_2 adsorption and desorption isotherms. Figure S2: Reduced sample TEM images. Figure S3: Spent sample Ni 2p XPS spectra. Figure S4: C1s XPS spectra. Figure S5: Catalyst activity and selectivity stability tests. Figure S6: Proposed catalytic mechanism on K-substituted LNO samples. Figure S7: Spent sample K 2p XPS spectra. Figure S8: Process flow diagram of reactor setup used for the catalytic evaluation.

Author Contributions: C.T. undertook the experiments, analysed the results, and wrote the manuscript. H.A., J.S., and R.A. conceived the experiments and assisted with the results analyses. Y.W., C.Y.T., and, R.J.W. assisted with experimental work, results analyses, and writing the manuscript. All authors have read and agreed to the published version of the manuscript.

Funding: This research was funded by the Australian Research Council under the Laureate Fellowship Scheme [FL140100081].

Acknowledgments: The authors acknowledge the use of facilities within the UNSW Mark Wainwright Analytical Centre.

Conflicts of Interest: The authors declare no conflict of interest.

References

1. Aziz, M.A.A.; Jalil, A.A.; Triwahyono, S.; Ahmad, A. CO_2 methanation over heterogeneous catalysts: Recent progress and future prospects. *Green Chem.* **2015**, *17*, 2647–2663. [CrossRef]
2. Song, C. Global challenges and strategies for control, conversion and utilization of CO_2 for sustainable development involving energy, catalysis, adsorption and chemical processing. *Catal. Today* **2006**, *115*, 2–32. [CrossRef]
3. Hunt, A.J.; Sin, E.H.K.; Marriott, R.; Clark, J.H. Generation, capture, and utilization of industrial carbon dioxide. *ChemSusChem* **2010**, *3*, 306–322. [CrossRef] [PubMed]
4. Centi, G.; Perathoner, S. Opportunities and prospects in the chemical recycling of carbon dioxide to fuels. *Catal. Today* **2009**, *148*, 191–205. [CrossRef]
5. Daza, Y.A.; Kuhn, J.N. CO_2 conversion by reverse water gas shift catalysis: Comparison of catalysts, mechanisms and their consequences for CO_2 conversion to liquid fuels. *RSC Adv.* **2016**, *6*, 49675–49691. [CrossRef]
6. Xu, J.; Froment, G.F. Methane steam reforming, methanation and water-gas shift: I. Intrinsic kinetics. *AIChE J.* **1989**, *35*, 88–96. [CrossRef]
7. Ohnishi, R.; Liu, S.; Dong, Q.; Wang, L.; Ichikawa, M. Catalytic dehydrocondensation of methane with CO and CO_2 toward benzene and naphthalene on Mo/HZSM-5 and Fe/Co-Modified Mo/HZSM-5. *J. Catal.* **1999**, *182*, 92–103. [CrossRef]
8. Wang, L.; Ohnishi, R.; Ichikawa, M. Selective dehydroaromatization of methane toward benzene on Re/HZSM-5 catalysts and effects of CO/CO_2 Addition. *J. Catal.* **2000**, *190*, 276–283. [CrossRef]
9. Liu, Z.; Nutt, M.A.; Iglesia, E. The effects of CO_2, CO and H_2 Co-reactants on methane reactions catalyzed by Mo/H-ZSM-5. *Catal. Lett.* **2002**, *81*, 271–279. [CrossRef]
10. Arandiyan, H.; Wang, Y.; Scott, J.; Mesgari, S.; Dai, H.; Amal, R. In situ exsolution of bimetallic Rh–Ni nanoalloys: A highly efficient catalyst for CO_2 methanation. *ACS Appl. Mater. Interfaces* **2018**, *10*, 16352–16357. [CrossRef]
11. Arandiyan, H.; Wang, Y.; Sun, H.; Rezaei, M.; Dai, H. Ordered meso- and macroporous perovskite oxide catalysts for emerging applications. *Chem. Commun.* **2018**, *54*, 6484–6502. [CrossRef] [PubMed]
12. Kim, C.H.; Qi, G.; Dahlberg, K.; Li, W. Strontium-doped perovskites rival platinum catalysts for treating NOx in simulated diesel exhaust. *Science* **2010**, *327*, 1624–1627. [CrossRef] [PubMed]

13. Wang, Y.; Arandiyan, H.; Scott, J.; Akia, M.; Dai, H.; Deng, J.; Aguey-Zinsou, K.-F.; Amal, R. High performance Au–Pd supported on 3D hybrid strontium-substituted lanthanum manganite perovskite catalyst for methane combustion. *ACS Catal.* **2016**, *6*, 6935–6947. [CrossRef]
14. Fukuhara, C.; Hayakawa, K.; Suzuki, Y.; Kawasaki, W.; Watanabe, R. A novel nickel-based structured catalyst for CO_2 methanation: A honeycomb-type Ni/CeO_2 catalyst to transform greenhouse gas into useful resources. *Appl. Catal. A* **2017**, *532*, 12–18. [CrossRef]
15. Ashok, J.; Ang, M.L.; Kawi, S. Enhanced activity of CO_2 methanation over Ni/CeO_2-ZrO_2 catalysts: Influence of preparation methods. *Catal. Today* **2017**, *281*, 304–311. [CrossRef]
16. Abdolrahmani, M.; Parvari, M.; Habibpoor, M. Effect of copper substitution and preparation methods on the $LaMnO_{3\pm\delta}$ structure and catalysis of methane combustion and CO oxidation. *Chin. J. Catal.* **2010**, *31*, 394–403. [CrossRef]
17. Singh, S.; Zubenko, D.; Rosen, B.A. Influence of $LaNiO_3$ shape on its solid-phase crystallization into coke-free reforming catalysts. *ACS Catal.* **2016**, *6*, 4199–4205. [CrossRef]
18. Fierro, J.L.G.; Tascón, J.M.D.; Tejuca, L.G. Surface properties of $LaNiO_3$: Kinetic studies of reduction and of oxygen adsorption. *J. Catal.* **1985**, *93*, 83–91. [CrossRef]
19. Weatherbee, G.D.; Bartholomew, C.H. Hydrogenation of CO_2 on group VIII metals: II. Kinetics and mechanism of CO_2 hydrogenation on nickel. *J. Catal.* **1982**, *77*, 460–472. [CrossRef]
20. Lin, W.; Stocker, K.M.; Schatz, G.C. Mechanisms of hydrogen-assisted CO_2 reduction on nickel. *J. Am. Chem. Soc.* **2017**, *139*, 4663–4666. [CrossRef]
21. Ding, Y.; Wang, S.; Zhang, L.; Chen, Z.; Wang, M.; Wang, S. A facile method to promote $LaMnO_3$ perovskite catalyst for combustion of methane. *Catal. Commun.* **2017**, *97*, 88–92. [CrossRef]
22. Chen, D.; Chen, C.; Baiyee, Z.M.; Shao, Z.; Ciucci, F. Nonstoichiometric oxides as low-cost and highly-efficient oxygen reduction/evolution catalysts for low-temperature electrochemical devices. *Chem. Rev.* **2015**, *115*, 9869–9921. [CrossRef] [PubMed]
23. Liang, B.; Duan, H.; Su, X.; Chen, X.; Huang, Y.; Chen, X.; Delgado, J.J.; Zhang, T. Promoting role of potassium in the reverse water gas shift reaction on Pt/mullite catalyst. *Catal. Today* **2017**, *281 Pt 2*, 319–326. [CrossRef]
24. Yang, X.L.; Su, X.; Chen, X.D.; Duan, H.M.; Liang, B.L.; Liu, Q.G.; Liu, X.Y.; Ren, Y.J.; Huang, Y.Q.; Zhang, T. Promotion effects of potassium on the activity and selectivity of Pt/zeolite catalysts for reverse water gas shift reaction. *Appl. Catal. B* **2017**, *216*, 95–105. [CrossRef]
25. Amoyal, M.; Vidruk-Nehemya, R.; Landau, M.V.; Herskowitz, M. Effect of potassium on the active phases of Fe catalysts for carbon dioxide conversion to liquid fuels through hydrogenation. *J. Catal.* **2017**, *348*, 29–39. [CrossRef]
26. Bansode, A.; Tidona, B.; von Rohr, P.R.; Urakawa, A. Impact of K and Ba promoters on CO_2 hydrogenation over Cu/Al_2O_3 catalysts at high pressure. *Catal. Sci. Technol.* **2013**, *3*, 767–778. [CrossRef]
27. Xu, J.; Liu, J.; Zhao, Z.; Xu, C.; Zheng, J.; Duan, A.; Jiang, G. Easy synthesis of three-dimensionally ordered macroporous $La_{1-x}K_xCoO_3$ catalysts and their high activities for the catalytic combustion of soot. *J. Catal.* **2011**, *282*, 1–12. [CrossRef]
28. Feng, N.; Chen, C.; Meng, J.; Wu, Y.; Liu, G.; Wang, L.; Wan, H.; Guan, G. Facile synthesis of three-dimensionally ordered macroporous silicon-doped $La_{0.8}K_{0.2}CoO_3$ perovskite catalysts for soot combustion. *Catal. Sci. Technol.* **2016**, *6*, 7718–7728. [CrossRef]
29. de Lima, S.M.; da Silva, A.M.; da Costa, L.O.O.; Assaf, J.M.; Mattos, L.V.; Sarkari, R.; Venugopal, A.; Noronha, F.B. Hydrogen production through oxidative steam reforming of ethanol over Ni-based catalysts derived from $La_{1-x}Ce_xNiO_3$ perovskite-type oxides. *Appl. Catal. B* **2012**, *121*, 1–9. [CrossRef]
30. Silva, P.P.; Ferreira, R.A.R.; Noronha, F.B.; Hori, C.E. Hydrogen production from steam and oxidative steam reforming of liquefied petroleum gas over cerium and strontium doped $LaNiO_3$ catalysts. *Catal. Today* **2017**, *289*, 211–221. [CrossRef]
31. Seaton, N.A. Determination of the connectivity of porous solids from nitrogen sorption measurements. *Chem. Eng. Sci.* **1991**, *46*, 1895–1909. [CrossRef]
32. Zhen, W.; Li, B.; Lu, G.; Ma, J. Enhancing catalytic activity and stability for CO_2 methanation on Ni-Ru/Al_2O_3 via modulating impregnation sequence and controlling surface active species. *RSC Adv.* **2014**, *4*, 16472–16479. [CrossRef]

33. Maneerung, T.; Hidajat, K.; Kawi, S. K-doped LaNiO₃ perovskite for high-temperature water-gas shift of reformate gas: Role of potassium on suppressing methanation. *Int. J. Hydrogen Energy* **2017**, *42*, 9840–9857. [CrossRef]
34. Miyakoshi, A.; Ueno, A.; Ichikawa, M. XPS and TPD characterization of manganese-substituted iron–potassium oxide catalysts which are selective for dehydrogenation of ethylbenzene into styrene. *Appl. Catal. A* **2001**, *219*, 249–258. [CrossRef]
35. Sawyer, R.; Nesbitt, H.W.; Secco, R.A. High resolution X-ray Photoelectron Spectroscopy (XPS) study of K₂O–SiO₂ glasses: Evidence for three types of O and at least two types of Si. *J. Non-Cryst. Solids* **2012**, *358*, 290–302. [CrossRef]
36. Misra, D.; Kundu, T.K. Transport properties and metal–insulator transition in oxygen deficient LaNiO₃: A density functional theory study. *Mater. Res. Express* **2016**, *3*, 095701. [CrossRef]
37. Mickevičius, S.; Grebinskij, S.; Bondarenka, V.; Vengalis, B.; Šliužienė, K.; Orlowski, B.A.; Osinniy, V.; Drube, W. Investigation of epitaxial LaNiO₃₋ₓ thin films by high-energy XPS. *J. Alloys Compd.* **2006**, *423*, 107–111. [CrossRef]
38. Kesavan, J.K.; Luisetto, I.; Tuti, S.; Meneghini, C.; Iucci, G.; Battocchio, C.; Mobilio, S.; Casciardi, S.; Sisto, R. Nickel supported on YSZ: The effect of Ni particle size on the catalytic activity for CO₂ methanation. *J. CO2 Util.* **2018**, *23*, 200–211. [CrossRef]
39. Su, X.; Yang, X.; Zhao, B.; Huang, Y. Designing of highly selective and high-temperature endurable RWGS heterogeneous catalysts: Recent advances and the future directions. *J. Energy Chem.* **2017**, *26*, 854–867. [CrossRef]
40. Sun, F.-m.; Yan, C.-f.; Wang, Z.-d.; Guo, C.-q.; Huang, S.-l. Ni/Ce–Zr–O catalyst for high CO₂ conversion during reverse water gas shift reaction (RWGS). *Int. J. Hydrogen Energy* **2015**, *40*, 15985–15993. [CrossRef]
41. Tada, S.; Shimizu, T.; Kameyama, H.; Haneda, T.; Kikuchi, R. Ni/CeO₂ catalysts with high CO₂ methanation activity and high CH₄ selectivity at low temperatures. *Int. J. Hydrogen Energy* **2012**, *37*, 5527–5531. [CrossRef]
42. Ullah, S.; Lovell, E.C.; Wong, R.J.; Tan, T.H.; Scott, J.A.; Amal, R. Light enhanced CO₂ reduction to CH₄ using non-precious transition metal catalysts. *ACS Sustain. Chem. Eng.* **2020**, *8*, 5056–5066. [CrossRef]
43. García-García, I.; Lovell, E.C.; Wong, R.J.; Barrio, V.L.; Scott, J.; Cambra, J.F.; Amal, R. Silver-Based Plasmonic Catalysts for Carbon Dioxide Reduction. *ACS Sustain. Chem. Eng.* **2020**, *8*, 1879–1887. [CrossRef]
44. Jantarang, S.; Lovell, E.C.; Tan, T.H.; Scott, J.; Amal, R. Role of support in photothermal carbon dioxide hydrogenation catalysed by Ni/CeₓTiᵧO₂. *Prog. Nat. Sci. Mat. Int.* **2018**, *28*, 168–177. [CrossRef]
45. Tejuca, L.G.; Fierro, J.L.G. XPS and TPD probe techniques for the study of LaNiO₃ perovskite oxide. *Thermochim. Acta* **1989**, *147*, 361–375. [CrossRef]
46. Day, J.P.; Pearson, R.G.; Basolo, F. Kinetics and mechanism of the thermal decomposition of nickel tetracarbonyl. *J. Am. Chem. Soc.* **1968**, *90*, 6933–6938. [CrossRef]
47. Lehman, R.L.; Gentry, J.S.; Glumac, N.G. Thermal stability of potassium carbonate near its melting point. *Thermochim. Acta* **1998**, *316*, 1–9. [CrossRef]
48. Zagli, A.E.; Falconer, J.L.; Keenan, C.A. Methanation on supported nickel catalysts using temperature programmed heating. *J. Catal.* **1979**, *56*, 453–467. [CrossRef]
49. Peebles, D.E.; Goodman, D.W.; White, J.M. Methanation of carbon dioxide on nickel(100) and the effects of surface modifiers. *J. Phys. Chem.* **1983**, *87*, 4378–4387. [CrossRef]
50. Dwivedi, N.; Yeo, R.J.; Satyanarayana, N.; Kundu, S.; Tripathy, S.; Bhatia, C.S. Understanding the role of nitrogen in plasma-assisted surface modification of magnetic recording media with and without ultrathin carbon overcoats. *Sci. Rep.* **2015**, *5*, 7772. [CrossRef]
51. Glatzel, P.; Singh, J.; Kvashnina, K.O.; van Bokhoven, J.A. In Situ Characterization of the 5d Density of States of Pt Nanoparticles upon Adsorption of CO. *J. Am. Chem. Soc.* **2010**, *132*, 2555–2557. [CrossRef]
52. Chen, C.-S.; Cheng, W.-H.; Lin, S.-S. Mechanism of CO formation in reverse water–gas shift reaction over Cu/Al₂O₃ catalyst. *Catal. Lett.* **2000**, *68*, 45–48. [CrossRef]
53. Wong Roong, J.; Tsounis, C.; Scott, J.; Low Gary, K.C.; Amal, R. Promoting catalytic oxygen activation by localized surface plasmon resonance: Effect of visible light pre-treatment and bimetallic interactions. *ChemCatChem* **2017**, *10*, 287–295. [CrossRef]

54. Wong, R.J.; Scott, J.; Kappen, P.; Low, G.K.C.; Hart, J.N.; Amal, R. Enhancing bimetallic synergy with light: The effect of UV light pre-treatment on catalytic oxygen activation by bimetallic Au–Pt nanoparticles on a TiO$_2$ support. *Catal. Sci. Technol.* **2017**, *7*, 4792–4805. [CrossRef]
55. Irusta, S.; Pina, M.P.; Menéndez, M.; Santamaría, J. Catalytic combustion of volatile organic compounds over La-based perovskites. *J. Catal.* **1998**, *179*, 400–412. [CrossRef]

© 2020 by the authors. Licensee MDPI, Basel, Switzerland. This article is an open access article distributed under the terms and conditions of the Creative Commons Attribution (CC BY) license (http://creativecommons.org/licenses/by/4.0/).

MDPI
St. Alban-Anlage 66
4052 Basel
Switzerland
Tel. +41 61 683 77 34
Fax +41 61 302 89 18
www.mdpi.com

Catalysts Editorial Office
E-mail: catalysts@mdpi.com
www.mdpi.com/journal/catalysts

www.ingramcontent.com/pod-product-compliance
Lightning Source LLC
LaVergne TN
LVHW070652100526
838202LV00013B/943

*9 7 8 3 0 3 6 5 3 9 7 8 2 *